D0518214

100 OF THE MOST EFFECTIVE WAYS TO

SUPERCHARGE YOUR METABOLISM

100 OF THE MOST EFFECTIVE WAYS TO
SUPERCHARGE YOUR METABOLISM

Get Your Body To Burn More Fat And Calories—
Safely, Easily, And Effectively

MOST METHODS TAKE 10 MINUTES OR LESS!

CYNTHIA PHILLIPS, PH.D.
Pierre Manfroy, M.D., and Shana Priwer

CRESTLINE

Inspiring | Educating | Creating | Entertaining

Brimming with creative inspiration, how-to projects, and useful information to enrich your everyday life, Quarto Knows is a favorite destination for those pursuing their interests and passions. Visit our site and dig deeper with our books into your area of interest: Quarto Creates, Quarto Cooks, Quarto Homes, Quarto Lives, Quarto Drives, Quarto Explores, Quarto Gifts, or Quarto Kids.

Text © 2009 Cynthia Phillips, Ph.D., and Shana Priwer

This edition published in 2018 by Crestline,
an imprint of The Quarto Group
142 West 36th Street, 4th Floor
New York, NY 10018 USA
T (212) 779-4972 F (212) 779-6058
www.QuartoKnows.com

First published in the USA in 2009 by Fair Winds Press,
an imprint of the Quarto Group, 100 Cummings Center,
Suite 265D, Beverly, MA 01915-6101

All rights reserved. No part of this book may be reproduced in any form without written permission of the copyright owners. All images in this book have been reproduced with the knowledge and prior consent of the artists concerned, and no responsibility is accepted by producer, publisher, or printer for any infringement of copyright or otherwise, arising from the contents of this publication. Every effort has been made to ensure that credits accurately comply with information supplied. We apologize for any inaccuracies that may have occurred and will resolve inaccurate or missing information in a subsequent reprinting of the book.

10 9 8 7 6 5 4 3 2 1

Crestline titles are also available at discount for retail, wholesale, promotional, and bulk purchase. For details, contact the Special Sales Manager by email at specialsales@quarto.com or by mail at The Quarto Group, Attn: Special Sales Manager, 401 Second Avenue North, Suite 310, Minneapolis, MN 55401, USA.

ISBN-13: 978-0-7858-3592-9

Printed and bound in China

Cover and book design by Fair Winds Press
Images: Istockphoto.com: pp. 17, 23, 25, 29, 31, 44, 48, 57, 61, 62, 65, 67, 71, 79, 83, 87, 89, 95, 101, 103, 105, 106, 111, 115, 120, 125, 129, 130, 139, 143, 147, 151, 162, 185, 189, 191, 195, 197, 199, 203, 213, 217, 219, 221, 231
Fotolia.com: pp. 5, 26, 35, 37, 39, 75, 77, 85, 91, 99, 113, 123, 127, 133, 135, 155, 159, 161, 165, 167, 173, 177, 179, 183, 225
Jupiterimages.com: pp. 13, 15, 18, 33, 43, 51, 119, 209, 211, 227

The information in this book is for educational purposes only. It is not intended to replace the advice of a physician or medical practitioner. Please see your health-care provider before beginning any new health program.

Contents

PART III

PART IV

PART V

PART VI

PART VII

Holistic Healing: Body-Mind
Techniques to Boost Your Metabolism . . . 204

FOREWORD

In my clinical practice, I work with many patients to help them manage their weight. Although success rates vary, generally the story goes something like this: A patient comes in for an initial office visit, enthusiastic about the idea of losing weight. We create a plan based on small but (hopefully) sustainable changes in the patient's diet, activity level, and overall lifestyle. We review and agree upon the plan. The patient is committed to try to stick to the plan.

Weeks go by, and weight falls away. The patient is happy, and has a lot of positive energy and momentum. Weeks to months later, the patient has lost a significant amount of weight and graduates from the weight-management program—but he or she leaves my office with a pledge to keep up the good work. Months later, perhaps at the annual checkup, we often find ourselves back at square one. The patient has regained some or most of the weight lost previously, perhaps even put on additional weight. This scenario is frustrating for everyone, including me. Studies validate my clinical observation: It is very difficult to keep weight off.

Why is it so hard to permanently shed weight? The answer is not a simple one because many factors are involved. Genetics play a clear role: Grandma is overweight, Mom is too, and Daughter is rapidly following the path to obesity. We also should not underestimate the influence of environment in the equation. If one person is trying to lose weight but is surrounded by a family of junk-food eaters, the task becomes exponentially harder. Economics may contribute as well. Many inexpensive foods and offerings at fast-food restaurants are high in calories, fat, and salt.

Metabolism is key to losing weight and keeping it off. A major reason why people have a hard time losing weight is that the body's natural response to weight loss is to lower its own metabolism. This phenomenon is an evolutionary safety net against starvation. Many years ago when food was less abundant than it is today, humans might have gone days or even weeks without food. During those times, the body needed defenses against starvation. When our ancestors were short on food and began losing weight, their bodies slowed down their metabolism to slow weight loss.

Today, we are left with a genetic makeup that retains this defense mechanism. Our bodies are disinclined to lose weight voluntarily. Indeed, your basal metabolic rate (BMR), or the rate at which you burn energy while doing absolutely nothing, is based largely on your weight. The lower your weight, the lower your BMR. As you lose weight, your metabolism declines, making it harder and harder to keep off the fat.

How can you fight this tendency by raising your metabolism? Is there a way to counteract your body's natural response and rev up your calorie-burning capabilities even when you're losing weight? In this book, authors Cynthia Phillips, Ph.D., and Shana Priwer have devised a balanced, fair approach to the many claimed "metabolism boosters" of the world. It is not easy to untangle today's science from propaganda, nor to keep up with all the messages that bombard us. If you peruse the Internet, for example, you will find many products that make assertions about raising metabolism. Small, poorly designed studies can be easily misinterpreted as fact. However, you'll also see many excellent websites that help the average reader sort through some of those claims, notably the Mayo Clinic patient website (www.mayoclinic.com) and the Linus Pauling Institute on micronutrient research website (http://lpi.oregonstate.edu/infocenter).

Phillips and Priwer wade through the fiction and get to the facts on metabolism boosters. The points in this book represent tips based on reasonable science, not unsubstantiated claims. The result is an honest approach to what can be a confusing topic. Readers who want to lose weight, keep it off, or just live a healthier lifestyle will find plenty of helpful information in this book. One hundred different suggestions are included—something for everyone.

Remember to consult your physician before making major dietary or lifestyle changes, and keep up to date on the latest research as best you can. That said, I wish you luck in putting these suggestions into practice.

To your health!

—Pierre Manfroy, M.D.

INTRODUCTION
Knowledge Is Power:
Understanding Your Metabolism

Before you can understand exactly why boosting your metabolism is vital to losing weight and maintaining a healthy lifestyle, you need to know what metabolism really means. In medical jargon, "metabolism" refers to the chemical reactions that are necessary to sustain life. These reactions occur in every living organism and, interestingly, the basic set of reactions is similar across many species. The word metabolism comes from the Greek word *metabole*, which means "change"—your body changes food into energy.

The chemical processes of metabolism fit into two broad categories. In the first, known as catabolism, organic materials such as food are broken down to give energy to cells. The processes of the second category, anabolism, take that energy and use it to build important cellular components like nucleic acids and proteins. Metabolism follows certain chemical pathways, using a series of chemical reactions to transform an initial material into a final product.

Measuring Your Metabolism

The term "metabolic rate" is used to describe one of a number of measurements of metabolic activity. The standard measure is the BMR, or basal metabolic rate. Your BMR represents how much energy is needed to sustain your body in a complete resting state. BMR is a precise measurement, and usually taken while subjects are fasting, in a neutral temperature, completely still, and unaroused. While exercise and other activities can raise metabolism, age and other factors cause a decline in BMR.

A less precise measurement based on BMR is RMR, which measures your resting metabolic state under less stringent conditions than BMR. Finally, there is TMR, or total metabolic rate, also referred to as TEE, or total energy expenditure. Your TMR is your BMR plus the calories you burn during the day with all your physical activity, plus the energy spent digesting the food you eat (also called diet-induced thermogenesis).

This book explores ways to boost all of these metabolic rates. Some of the 100 metabolic boosters aim to elevate your BMR, mainly by increasing your muscle mass—muscle is a metabolically active tissue that burns a higher rate of calories than other tissue, even when inactive. Others are intended to raise your physical activity, and ultimately your TMR/

TEE. Yet others are designed to help you get the most energy expenditure out of the food you eat, increasing your diet-induced thermogenesis and ultimately your TMR/TEE.

Basal Metabolic Rate (BMR)

Although metabolism may seem like an abstract concept, you can easily calculate your BMR. In clinical tests, scientists place the subject in a controlled environment and measure how much oxygen the person consumes and how much carbon dioxide he or she gives off. While you can't do that at home, you can estimate your BMR using some basic information about yourself.

The BMR takes account of weight, height, age, and gender, so there are two formulas, one for men and one for women, and each has a different formula in English (Imperial) units and in metric units, as shown below.

The "average female"—thirty-five years old, five feet six inches (168 cm) tall, and weighing 140 pounds (64 kg)—has a BMR of 1410. Ten years later, assuming her weight has remained constant, that same female will have a BMR of 1363. A female who is thirty-five years old,

five feet six inches (168 cm) tall, and weighs 200 pounds (91 kg) has a BMR of 1671. The BMR for a thirty-five-year-old man who is six feet (183 cm) tall and weighs 180 pounds (82 kg) will be 1864. Ten years later, his BMR will have declined to 1796.

Bear in mind that the BMR does not take into account a person's percentage of muscle mass versus body fat. Thus, may not be as useful a tool for the elite athlete with a very high percentage of muscle mass as it is for people in the middle range, nor will it be quite as useful for someone who is extremely obese and has a larger-than-usual amount of body fat.

ENGLISH (IMPERIAL) FORMULAS

Male
BMR = 66 + (6.23 x weight in pounds) + (12.7 x height in inches) − (6.8 x age in years)

Female
BMR = 655 + (4.35 x weight in pounds) + (4.7 x height in inches) − (4.7 x age in years)

METRIC FORMULAS

Male
BMR = 66 + (13.7 x weight in kg) + (5 x height in cm) − (6.8 x age in years)

Female
BMR = 655 + (9.6 x weight in kg) + (1.8 x height in cm) − (4.7 x age in years)

Body Mass Index (BMI)

Another useful tool for understanding metabolism is your BMI, or body mass index. Developed in the nineteenth century, this index wasn't widely adopted until the 1980s. It uses a weight-per-height formula to help you assess whether you are underweight, healthy, overweight, or obese. The formulas are different for English and metric units, as shown at right.

So, if you are using English units, multiply your body weight in pounds by 703 and then divide it by the square of your height in inches (inches x inches). For example, if you weigh 140 pounds (64 kg) and stand five feet six inches (168 cm) tall, your BMI is 140 times 703 (98,420) divided by 66 times 66 (4,356); that is 22.6, which is usually rounded up to 23. If you struggle with the math, don't worry; there are plenty of BMI calculators available online and from health-care professionals.

But what does your BMI number mean? The value falls into one of four categories, as shown here:

BMI CATEGORIES	
18.5 or less	Underweight
18.5 to 24.9	Normal
25 to 29.9	Overweight
30 and over	Obese

ENGLISH (IMPERIAL) FORMULAS
$BMI = weight\ (pounds) \times 703\ /\ height\ (inches)^2$

METRIC FORMULAS
$BMI = weight\ (kg)\ /\ height\ (m)^2$

So the BMI lets you work out your current body-weight status and helps you assess how far you have to go to get to a normal body weight. Note, however, that the BMI does not take into account the fact that muscle mass weighs more than body fat. As a result, some individuals with high muscle mass and low body fat may be inappropriately labeled overweight or obese on the BMI chart. Similarly, BMI might not be entirely valid for older adults, who may have lost bone or muscle mass as they have aged. And it is not the best tool for measuring young children, for whom height-and-weight trends are generally more reliable indicators of health.

Weight Loss and Metabolism

Once you understand how your BMR is calculated, it's easy to understand the impact of weight loss on this rate. Let's take another look at our "average" female, thirty-five years old, five feet six inches tall (168 cm), and weighing 140 pounds (64 kg). She has a BMR of 1410. If she gains sixty pounds (27 kg), she'll have a BMR of 1671. At 240 pounds (109 kg), her BMR is 1845.

Does this mean that gaining weight is a legitimate way to raise metabolism? Technically yes, but practically no. What it indicates is that people who weigh more require more calories to sustain that weight. Remember, BMR is just part of the picture when it comes to total energy expenditure on a daily basis. Losing weight may decrease your BMR, but your TMR (total metabolic rate) has likely increased due to the increased activities and healthier lifestyle you've adopted to lose that weight.

Let's also take a look at the BMIs for our example female. At 240 pounds (109 kg) she has a BMI of 39, placing her well within the obese range. At 200 pounds (91 kg) she has a BMI of 32, which still flags her as obese. At her target weight of 140 pounds (64 kg), her BMI is 23, well within the "normal" range.

Losing weight and raising metabolism are codependent activities. Shedding weight effectively involves raising metabolic rate, because a faster metabolism allows you to burn more calories per day. Adding more metabolically active muscle tissue to your body raises your BMR. The aerobic and anaerobic exercise that contribute to building muscle and losing fat also burn calories, which contributes to weight loss and an increased TMR/TEE.

Expecting success with losing weight and raising metabolism is one of the best ways to realize success. However, things don't always go according to plan, and most of us know that losing weight can be a challenge. Remember that the essential formula for weight loss is, again, a numbers game: More calories must be burned than are consumed. But many factors can impede your progress.

Variations in Metabolic Rate

Like most things in life, metabolic rate is not constant. It changes year to year, day to day, hour to hour, and often more frequently. Just as there are diurnal (daytime–nighttime) variations in the blood levels of hormones and other chemicals, there are also similar variations in metabolic rate. A wide range of factors can influence a person's metabolism. These include, but are not limited to, gender, weight, age, illness, diet, emotional health, genetics, body-fat percentage, body temperature, fitness level, and even ambient air temperature. Most of these can change, and will change during the course of your life.

Change itself is not necessarily bad for your metabolism. On the contrary, changing something like an exercise routine can actually stimulate metabolic rate, which is sensitive to changes in blood oxygen levels. However, the extent to which these factors vary may be beyond our conscious control—or so we think. Consider aging, for example, a process that happens to every living thing. As you get older, your BMR tends to decline due to changes in your body's cells and decreased muscle mass. (To counteract this, you need to incorporate additional weight-bearing exercises into your fitness routine.)

Gender is another metabolic determinant. Generally speaking, men have more muscle mass and less body fat than women because of their different hormonal allocations, so their BMRs tend to be higher. Genetics also play a role and some people are simply born with faster metabolisms than others.

Many daily habits have an impact on metabolism. Exercise, which burns calories and builds muscle mass, has the net effect of raising metabolism. Whether it's aerobic exercise, yoga, or weight training, all forms of physical activity get metabolism moving in the right direction. Other, more cyclical, events contribute to a fluctuating

metabolism. Catching a bad cold can increase metabolic rate because when you have a fever, your body temperature increases, which in turn raises the metabolic rate. Similarly, when you are exposed to cold weather, your body works harder to produce the heat necessary to keep you warm, and that requires a rise in metabolic rate.

Mood Matters

Emotional eating: We've all been there. A negative emotion (anger, frustration, tension, or even boredom) triggers us to eat, even when we're not physiologically hungry. Emotional hunger has distinct traits that set it apart from real hunger. The emotional variety tends to develop quickly (whereas physiological hunger comes on more slowly), and is often accompanied by cravings for one or more particular foods, and it can appear even if you've eaten recently. Frequently, emotional eating is followed by feelings of guilt or embarrassment.

Such eating can provide a welcome distraction from a difficult emotional issue, and "feel-good" comfort foods can prompt the release of endorphins that help us feel better. Once their effects wear off, though, it's back to the original problem—and, in many cases, a cycle of eating in response to emotional triggers can develop.

The metabolic impact of emotional eating can be disastrous. Consuming more calories than you burn off can lead to weight gain; eating high-fat, high-sugar comfort foods can encourage fat storage. The more overweight you are, the more sluggish you are likely to feel. And with less energy to exercise and boost your metabolic rate, it's easy to slip into a spiral toward a slower overall metabolism.

Emotions can also stimulate hormonal changes in the body that affect metabolic rate. Hormones are produced by the body's endocrine system, which also plays a large role in metabolism. Adrenaline, for example, is generated in response to a "fight or flight" survival instinct, and gives the muscles more blood in order to perform at a more intense pace. This in turn can affect metabolic rate.

The Nitty-Gritty on Burning Fat

Many of us eat for social, emotional, or other reasons. We may eat to share good feelings or to hide bad ones. Our motives may have little to do with the primary purpose of consuming food: to provide our bodies with fuel, from which we make energy. Unfortunately, if we supply our bodies with more fuel than we need, some of that extra fuel will be likely stored as fat.

Different kinds of fat are found in the human body. Composed of fatty acids and combined with glycerol, triglycerides are common in the bloodstream and are one of the main ways in which fat is transported to different parts of the body. Cholesterol has some structural similarities to triglycerides, but different functions. It is essential for the building of cell membranes, and helps produce acids and hormones; it therefore plays a vital role in making your metabolism work effectively.

Cholesterol is carried through the bloodstream by two kinds of what are known as lipoproteins: LDL (low-density lipoprotein) cholesterol, which is considered "bad" cholesterol because high levels can block arteries; and HDL (high-density lipoprotein) cholesterol, which helps remove LDL and is therefore considered "good" cholesterol. Like all types of fat, triglycerides and LDL cholesterol need to be kept within reasonable levels in order for you to look and feel healthy.

How is fat burned? When the body requires energy, triglycerides break down chemically, and the components are biochemically converted into energy-rich phosphate bonds, liberating energy that can be used for other cellular functions. Of the metabolic leftovers,

carbon dioxide is exhaled through the lungs, and hydrogen combines with oxygen molecules to create water, which is removed from the body via such functions as perspiration and urination.

If a person takes in more fuel than required, however, the leftover calories may be converted to fat. High triglyceride levels can be found in some individuals whose bodies are storing an abundance of calories as fat. A 2006 review article in the journal *Current Opinions in Cardiology* explained that a higher-than-average concentration of triglycerides in the blood is associated with a condition called "metabolic syndrome." This term refers to a group of conditions also including high blood pressure, high blood sugar, and low HDL cholesterol. Having these factors is thought to place a person at higher risk of diabetes, stroke, and heart disease.

Metabolism, Fat, and Muscle

Our bodies require fat, however—in appropriate amounts. Fat, carbohydrates, and protein are the body's basic building blocks. Each of these "fuel sources" serves specific needs. For example, high-intensity exercise usually depends on carbohydrates, whereas fat provides much of the fuel source for longer-duration exercise.

Regardless of the fuel source, the human body burns calories whether you're exercising or not. A person of average weight burns approximately a calorie a minute simply by sitting and breathing, which averages out to roughly 1,400 calories a day. More active people, and especially people who exercise throughout the day, will naturally burn far more calories.

Fat and metabolic rate have an intimate relationship. Metabolism includes the process of converting food into either energy to be used right away, or stored energy in the form of fat. A higher metabolism burns more calories and stores less fat, whereas a sluggish metabolism does the opposite.

Muscle is called "metabolically active" tissue because it places higher demands on your metabolism than fat does. This means that metabolism is higher when more muscle mass is present. Muscle burns more calories than fat, so if two people weigh the same but Person A has a high percentage of body fat and Person B has more lean body mass (such as muscle tissue), Person A's metabolism will (in the absence of other mitigating factors) be lower than that of Person B.

As you might expect, exercise and strength training are two of the best ways to reduce the size of the fat cells in your body. Diet is another component of this fat-burning strategy. The lesson for anyone trying to increase his or her BMR is a simple one: lose fat and gain muscle. Many of the points in the book are designed to help you do exactly that.

There are no quick fixes to burning fat, however. You cannot take a pill to burn it off while you sleep, or spread on a lotion that will melt it away while you lounge in front of the TV. For now, at least, we do know that self-regulation, aerobic exercise, strength training, and proper nutrition are the keys to burning fat and elevating metabolism.

Establish a Goal, Maintain the Right Attitude

We all have a tendency to compare ourselves to others, especially those we see on television and in movies. Actors, models, and athletes often present "ideal" images of the human body, which the rest of us are unlikely to achieve. However, good goals *are* necessary to lose weight and raise metabolism. Goals enable you to visualize your desired outcome, and being able to see yourself as healthy and confident contributes immeasurably to your ability to make your vision a reality.

Stepping on the scale can be intimidating when you're trying to lose weight. That measurement can determine so many things—how you feel about yourself, where you land on the obesity charts, and how much you allow yourself to eat. In our discussion of metabolism, we've presented a lot of numbers and numeric formulas. But when it comes to raising metabolism, playing solely by the numbers can be dangerous because all bodies are individual, and all bodies

respond differently to therapies, modalities, and habits. There is no one-size-fits-all approach that will result in a permanent rise in metabolic rate.

Each individual must take responsibility for developing his or her own plan for raising metabolism, losing weight, or becoming a healthier, happier human. Use the information presented in this book in the spirit in which it was intended. Pick and choose the methods that work for *you*. Creating a customized goal, and a customized means of achieving that goal, will set you up for success.

A Combination of Strategies

Nutrition, exercise, and self-regulation can loosely be considered the three main requirements for raising metabolism. No one method, conducted to the exclusion of the other two, will result in permanent lifestyle changes. Raising metabolism, and keeping it high for life, relies on a combination approach. That's because regulating metabolism involves many systems in the body. You need to consider the heart and lungs, endocrine system, digestive tract, and just about everything in between—in addition to the most important of all: the brain.

The entries in this book suggest specific techniques, means, and therapies for losing weight and raising metabolism.

We don't expect you to follow every single suggestion through to its logical conclusion. Some of the entries will resonate with particular individuals more than others. You'll naturally gravitate toward the ones that ring true for you.

This is not "just another diet book." The methods and approaches described herein are not a formula, prescription, recipe, or other set of instructions that must be followed to the letter, then forsaken once your goals have been met. Maintaining a finely tuned metabolism is a life-long endeavor—and this book is all about creating healthy, life-long habits. Don't expect to use these suggestions once or twice. To permanently change your habits and old ways of thinking, you'll probably need to practice the same points over and over, until they become second nature.

Finally, none of the information in this book is intended to be a substitute for consultation, diagnosis, and treatment from a medical professional, nor are any of the tips presented here to be used to diagnose or treat any disease. Never try to diagnose or treat yourself; consult your physician, specialists, and medical team members for this purpose.

PART I

Get Moving: Strength Training,
Flexibility Work, and Aerobics
for Increased Metabolism

Stretch Your Way to a Speedier Metabolism

Stretch often—especially if you don't have time for a full workout. According to a 2001 study that appeared in the *Scandinavian Journal of Medicine & Science in Sports*, moderate stretching can help improve your overall strength. That's good news for anyone who's trying to build muscle mass, which naturally raises your metabolism.

According to professional weight-loss consultant Anne Collins, you burn 122 calories with forty-five minutes of stretching. That adds up to 44,530 calories, or about thirteen pounds (6 kg) of body weight, in a year. Although that may not sound like a huge amount, stretching also energizes your body and boosts circulation, which in turn elevates your metabolism. The act of stretching stimulates blood flow to your muscles—as well as to other parts of the body—and increases energy production.

Stretching helps each component of a muscle relate smoothly to all the nearby muscles, and enhances their overall performance. It can be especially beneficial for older people who want to retain muscle mass and promote healthy metabolisms. An article published in *The Journals of Gerontology* in 2003 indicated that stretching exercises help to slow the process of age-related muscle wasting, increase the rate of muscle-protein synthesis, and improve muscle strength in senior citizens.

Stretch Throughout the Day

Do you spend entirely too much time sitting at a desk? Liven up your workday by taking periodic breaks to do some simple stretches—just a few minutes several times a day will help invigorate your system and boost your metabolism.

Quick and easy chair stretches require no special equipment—you don't even have to get out of your chair. Stretch your upper back by clasping the fingers of both hands together and extending your arms in front of your body. Relax your shoulders and stretch your neck by slowly leaning your head to the right, then to the left. To stretch your biceps hold both arms out at 90-degree angles to the body, palms facing ahead of you; slowly rotate your arms so that your palms point down toward the floor. Stretch your triceps by holding both arms straight over your head and clasping hands; bend your arms so both hands are behind your head, then raise your arms until they're fully extended over your head again. (You can enhance this stretch by holding a light weight, such as a bottle of water, in each hand.)

Easy Ways to Work the Legs

Here's an easy exercise that stretches your calves. You can do it at work, home, anyplace—all you need is a wall for balance. Stand about eighteen inches (45 cm) away from the wall and press your palms flat against the wall, keeping your arms straight. Slide your right foot back about eighteen inches (45 cm). Bend your left knee and lean forward until you feel a slight pulling sensation in the calf muscle. Release, and repeat on the opposite side.

If space allows, you can add some simple floor stretches to your repertoire. To stretch your hips, lie on your back and bring your right knee up to your chest. Hold the knee in place with your left hand and move it slightly to the left, until you feel a pulling sensation. Repeat on the opposite side.

Stretch your quadriceps by lying on your left side, with the left leg slightly bent. Slowly bring your right leg back until the ankle approaches the buttocks. Hold for a few moments before releasing, then repeat on the opposite side.

Listen to Your Body

How long should you hold each stretch? Listen to your body. Hold a position until you feel a slight pull in the muscle that's being stretched. If it hurts, the muscle was probably stretched too far or for too long. Go easier on that muscle group the next time. If your muscles are so relaxed that you feel no pulling, try holding the position for thirty to sixty seconds. Stretch smoothly and gently; avoid bouncing. Start slowly, and gradually increase the duration and number of stretches. Done safely, stretch breaks are a relaxing way to burn calories and boost your metabolism.

2

Take Weight Off with High-Intensity Weight Training

Building lean muscle mass is one of the time-tested, proven ways to increase resting metabolism. If you don't have a lot of time to exercise, but still want to add muscle, high-intensity weight training may be an ideal approach. A 1994 study published in the journal *Metabolism* found that test subjects who performed high-intensity workouts burned significantly more fat than those who participated in conventional regular-strength workouts.

High-intensity training involves working out more intensively than you would in a conventional weight-training program (▶3), but less often—maybe two or three times a week. You also perform fewer repetitions, but use more strength. For example, a conventional weight-lifting session might involve many repetitions using perhaps 70 to 75 percent of your strength. High-intensity lifting means you do fewer repetitions, but with heavier weights, so that you use 85 to 90 percent of your strength. Conventional weight-lifting sessions can last an hour or more; high-intensity weight-lifting sessions are usually much shorter. Weight lifters frequently take breaks of up to five minutes between sets, but high-intensity trainers will immediately get up and move to another exercise, working a different set of muscles, rather than resting between sets. This higher level of strength and activity lets you reap huge benefits. Although you are spending a shorter amount of time working out, your metabolism is enhanced even after you finish your workout, due to the addition of more lean muscle mass.

Build More Muscle, Burn More Fat

Lifting weights burns calories but, more importantly, it builds muscle. Exercising at a higher intensity burns even more calories. Combining weight lifting with moving at a faster, more intense pace allows you also to reap some of the benefits of a cardio workout, namely the fat-burning boost that comes from raising your heart rate and keeping it high for a period of time.

Another major benefit of a high-intensity workout, proponents say, is that you're focusing more on building muscle tissue, which raises metabolism and burns more fat than a normal weight-training session. According to an article published on former Mr. Universe Mike Mentzer's website, "The most important contribution that exercise makes to a fat-loss program is the maintenance of muscle tissue while fat is lost. Strength training is the only reliable method of maintaining muscle tissue."

The amount of strength training required to burn an extra 3,500 calories per week—that's one pound (0.45 kg)

of fat—is less than one hour per week. That's a significant metabolic boost for a relatively small investment of time.

Exhaust Your Muscles to Increase Strength

Maintain a high level of activity during a high-intensity weight-training session, and aim to increase the amount of weight you lift, no matter how slight, during each workout. By maximizing the sets you can achieve in a single session, you'll burn more calories and strengthen more muscle groups, both of which can help raise metabolism.

To avoid injury and soreness, make sure you are in reasonable condition before you try to lift any weights. If you have any concerns, ask your doctor's advice and obtain his or her approval before starting. Even then, a few sessions with a personal trainer can be incredibly helpful if you don't have any weight-lifting experience.

Use smooth, continuous movements while lifting. If you find that you have to rest too much between reps, change to a lighter weight that lets you maintain a consistent pace. The goal of high-intensity training is to exhaust your muscles, as long as you're physically able. A high-intensity training session is considered complete when your body feels ready to stop—not when the clock says it's time to go.

High-intensity strength training may or may not work for your particular body type, schedule, and other fitness goals, but its metabolic benefits may be reason enough to try it.

3 Build Muscles Slowly but Surely with Low-Intensity Resistance

If you are new to exercise or need to work out at a slower, more relaxed pace—for health reasons or personal preference—low-intensity weight training could be an ideal metabolism booster. A study published in 2006 in the *Scandinavian Journal of Medicine & Science in Sports* showed that low-intensity training improved blood lipid (fat) profiles and metabolic fitness.

Also called low-intensity resistance work, this type of weight training takes pretty much the opposite approach to high-intensity training. Whereas high-intensity training involves using heavier weights, fewer repetitions, and a quick workout pace, the idea behind low-intensity training is to do more lifting (say ten to fifteen repetitions) with lighter weights, until you can do all your repetitions without feeling exhausted. When you're able to achieve those repetitions easily, you increase the weight enough so that your muscles feel challenged again. Low-intensity resistance work involves using a variety of weights and machines for different muscle groups. Many people choose to work their arms one day and legs the next, for example. Consult with a personal trainer to determine your own limitations and goal weights.

Burn More Calories Safely

Although high-intensity lifting can give you the biggest bang for your buck, for many people the old "no pain, no gain" attitude can be self-defeating or even dangerous. For example, a 2002 University of Florida study indicates that older adults may obtain significant benefits from low-intensity strength training. The study found that for seniors this type of training not only firmed muscles, but also improved their overall physical endurance and aerobic power.

Reducing body fat and increasing lean muscle tissue is the key to raising metabolism, and low-intensity exercise may actually burn a higher percentage of calories *from fat* than high-intensity exercise does. However, as Jeremy Gentles, ObesityHelp's staff health and fitness expert points out, the *total* number of calories burned is more important than what percentage of these calories comes from fat. So, in order to maximize the fat-burning potential of added muscle tissue, experts suggest you challenge yourself with progressively heavier weights, even if you're staying well under your exertion level.

4 Burn More Calories by Doing More Sets

Regardless of whether you're doing high-intensity or low-intensity weight training (▶**2, 3**), you'll burn more calories if you do more sets. A "set" is the number of repetitions you perform—anywhere from two to twenty—before taking a rest break. A repetition (often called a "rep") is the act of pushing, pulling, or otherwise moving a weight through its range, for a given exercise.

Studies, such as a 2008 article published in the journal *Obesity*, confirm that when you're losing weight, weight lifting helps maintain muscle mass. The best weight-lifting methods, though, are the subject of contention. High-intensity and low-intensity weight training offer two different views on the philosophy behind strength training, and both have pluses and minuses.

If you're a beginner at weightlifting, consider the 1995 guidelines published by the American College of Sports Medicine, which recommend starting with between one and three sets of approximately ten repetitions per exercise; a 2003 article in the journal *Medicine & Science in Sports & Exercise* backs this up. As a general guideline, use weights that are light enough to let you successfully achieve your desired number of repetitions, but not so light that you don't feel challenged.

Adding Weight and Sets Burns More Calories

When you have gained some experience and your muscles are no longer being challenged by your current level of workout—you'll know because you can finish your sets without feeling any ache or "burn" afterward), you can steadily increase the weights you are moving. Professional trainers, such as ObesityHelp's staff health and fitness expert, Jeremy Gentles, explain that you will burn more calories exercising at a higher intensity, using heavier weights.

For example, a person who weighs 160 pounds (73 kg) will burn off about 215 calories during a thirty-minute low-intensity weight-lifting session. The same individual will burn 375 calories if he exercises for the same amount of time, but at a high intensity. However, high-intensity weight lifting isn't for everyone, and low-intensity training also burns calories, builds muscle mass, and stimulates metabolism. Either way, doing more reps and more sets results in more calories being burned.

A number of studies, such as the 2003 article in *Medicine & Science in Sports & Exercise* noted above, suggest you should increase both the weight and the number of sets you do, as you progress. Increasing your total volume of weight and sets will ultimately help you add more muscle tissue to your body, which results in elevating your metabolism even more.

When lifting weights to raise metabolism, give yourself forty-eight hours between sessions that work the same muscles; this much time is required for your muscles to recover and rebuild. And always listen to your body. If a weight feels too heavy, or a muscle, tissue, ligament, or any other part of your body is being moved in a painful way, stop immediately.

5 Burn Fat with Cardiovascular Exercise

Cardiovascular exercise is a proven way to burn body fat, especially when it is combined with strength training (▶2, 3). Also called aerobic exercise, cardio exercise, or simply "cardio," cardiovascular exercise means any type of exercise that, when done at a medium level of intensity, gets your heart rate up to somewhere between 65 and 80 percent of its maximum.

Running, jogging, stair climbing, rope jumping, cross-country skiing, and speed walking can all do this. If you suffer from knee or joint pain, try running or walking on an elliptical trainer, which eliminates most of the impact of your feet hitting the ground. Cycling, either on an outdoor or a stationary bicycle, generally requires fewer calories than running but still provides a great workout. Swimming offers an ideal, impact-free workout that burns plenty of calories as long as you keep up a quick pace. Rowing, or using a rowing machine, is another good low-impact choice, though it may not be suitable for those with shoulder, elbow, or lower-back pain.

Getting the Most out of Your Workout

Most heart-rate monitors and cardio-machine readouts show a "fat-burning zone," which is at about 60 to 70 percent of your maximum heart rate. Although studies such as one published in a 2002 article in the journal *Medicine & Science in Sports & Exercise* suggest that this may be an efficient heart rate for burning fat, basic physics and chemistry indicate that you'll burn more calories if you work out at a higher intensity, ideally 80 to 85 percent of your maximum heart rate. A 130-pound (59 kg) woman, for example, will burn about five calories per minute with her heart rate at 65 percent of maximum. She'll burn closer to seven calories per minute with her heart rate at 85 percent of maximum. In thirty minutes, she'll burn seventy-three or eighty-two calories, respectively. Performing at a lower intensity, however, this woman will burn 50 percent of those calories from fat, compared to less than 40 percent if she works out at a higher intensity.

One of the keys to getting the most from your cardio workout is to exercise for a sufficiently long time period. During the first fifteen minutes or so (depending on the intensity of your workout), glycogen supplies most of your energy. Once the body's glycogen stores decrease, however, you begin using fat for energy. Here's where most of the metabolic advantage comes into play. In addition to working muscles and burning calories, your body is also reducing its stored fat. Exercising for at least half an hour helps make sure that you burn both glycogen and fat stores. Additionally, because the glycogen stores in your muscles are lower after fasting overnight than they are after you've eaten a meal, some experts think exercising first thing in the morning is more effective for raising metabolism than exercising later in the day (▶63). Exercising at any time of day, however, will raise your metabolism while you're working out and for a couple of hours or so afterward.

6

Lose Weight Fast with Weighted Cardio Exercise

If you have a busy schedule, adding weights to your body while you work out can boost your metabolism while providing a more time-efficient cardio workout. Studies, such as one cited in a 2004 article published in the *Journal of the American Geriatrics Society*, indicate that wearing weighted devices like vests can help improve muscle capabilities. A 1980 article in *Track and Field Quarterly Review* showed significant gains in strength and ability from training with weighted vests and ankle weights.

Weighted Exercise Combines Cardio and Strength Training

When using this mode of exercise, you add weights to your body, then perform your chosen mode of cardiovascular exercise. Cardio exercise is necessary for losing weight; strength training helps build muscle. Both types of exercise are requirements for raising metabolism. Weighted cardio exercise combines the two types and gives you the best of both worlds. By weighing down specific parts of your body, you force your muscles to do extra work in order to keep up your usual pace, providing resistance training for the muscles involved. According to an article published in *Strength and Conditioning Journal* in 2000, resistance training improves strength, power, and endurance, as well as building lean body mass and reducing fat.

The additional work required from your body also burns more calories. According to Len Kravitz, Ph.D., in an article for the American Fitness Professionals and Associates (AFPA), resistance weight training ups the amount of energy you expend while exercising, which can promote weight and body fat loss. Furthermore, resistance training uses the larger muscle groups and increases your workout's total volume—the number of repetitions times the number of sets (▶4) times the amount of weight. Dr. Kravitz notes, "An impressive finding to highlight with resistance training is that the energy expenditure following the higher total volume workouts appears to be elevated, compared to other forms of exercise, and thus, further contributes to weight loss objectives."

Wearing Weights Makes Your Body Work Harder

Wearing a cardio weight vest—usually available in a range of weights from fifteen pounds (7 kg) to one hundred pounds (45 kg)—while walking, jogging, jumping rope, or boxing is one way to increase the intensity of a workout. Ankle and wrist weights bump up the strength-training effect of a fast-walking cardio workout. Wrist weights can help tone your arms while walking.

Experts say that using ankle weights while walking can also improve quadriceps and hamstring muscles significantly more than walking without them. Wearing a weighted belt while doing pushups and squats increases the intensity of the exercise as well as the number of calories burned.

Adding weights to a cardio workout thus increases metabolic benefits, but there can also be a downside. Weights place extra stress on joints. Therefore, experts often point out that you can run a greater risk of injury from adding weights to some types of cardio exercise. Don't use weights while running, for example, as this can significantly increase the impact on the joints in your knees, ankles, and feet. Consult with a sports therapist prior to beginning weighted cardio work, in order to mitigate these possibilities.

7 Build Stronger Muscles Using Free Weights

You can get a fantastic weight-training workout using free weights instead of, or in addition to, machines. Some sources suggest that your muscles are more fully engaged when you're working out with free weights because there's no machine to "help" any part of the process.

Strength studies comparing the benefits of free weights against machines, such as those cited in a 2000 article published in *Strength and Conditioning Journal*, show that "free weights consistently produce superior strength gains." As Doug Walker, an Indianapolis personal trainer and gym owner, explains, "Free weights involve more muscle fibers and more contractions, because you use stabilizing muscles not used when exercising on a machine; therefore free weights are more effective."

Increase Your Muscle Mass

Because you're working harder, you may actually build stronger muscles using free weights—and that means you'll have more endurance for a longer workout. The additional time spent exercising those muscles can translate into a higher resting metabolism. Free weights also allow you to do a number of different large muscle mass exercises, which require more energy than small muscle mass exercises. And strengthening more muscle groups can help raise your metabolism (▶2).

Expending more energy means that these exercises also help change body fat composition and metabolism, as shown by a 1991 study published in the journal *Sports Medicine*. Additionally, experts at the Mayo Clinic indicated in 2007 that using free weights develops more core support, as you're responsible for both lifting the weight and stabilizing your body simultaneously.

Get Muscle-Building Benefits with Resistance Bands

Novice weight lifters usually require a few sessions with someone knowledgeable before lifting weights by themselves. If you find that even the lowest weight in a free weight set is too difficult for you, you can start by using elastic resistance bands to gradually build up your muscle strength. These long elastic-rubber tubes offer weight-like resistance when you pull on them. Available in a range of lengths and tensions, they are ideal for beginners, senior citizens, and people who are recovering from injuries.

8

Sprint to Your Goal with Sprint Training

Interval or sprint training lets you burn fat and calories fast, without spending a lot of time at the gym. Studies, such as one published in 2006 in the *Journal of Applied Physiology*, show that you can burn a significant amount of fat doing interval training. Another 2005 study, published in the same journal, indicated that using sprint-training techniques improved exercise performance impressively after several weeks.

Sprint or interval training involves varying the pace of your workout by incorporating one- to two-minute bursts—or "intervals"—of intense activity. After each interval, you return to your normal pace for a period of three to five minutes, then add another interval. If you're running or exercising with a pedometer or distance counter,

you might sprint part of the way, for example. Try to keep the length of that sprint period consistent.

Changing Exercise Levels Challenges Your Body

Intervals are effective because they provide a challenge to your body. Research, such as a 1992 article published in the journal *Medicine & Science in Sports & Exercise*, says that if you always exercise at the same level—one within your comfort zone—your body adapts to the point where you may not even break a sweat. By measurably increasing the intensity of your activity, you will raise your heart rate and burn more calories, and help elevate your aerobic threshold (the point at which your body requires more oxygen and you start breathing harder and faster).

All exercise causes the body's metabolic rate to rise for a short period. However, evidence suggests that interval training may cause the metabolism to remain high for much longer. For example, a 1985 study published in the *American Journal of Clinical Nutrition* showed that if you exercise at a high level of intensity your metabolism can stay elevated for up to nine hours after exercising, whereas exercising at a very low level of intensity did not cause any longer-term metabolic gains.

Sprint Training Helps You Reap Metabolic Benefits More Quickly

Another benefit of interval training is that you don't need to work out for quite as long in order to reap the benefits of cardio exercise. A 2006 study in the *Journal of Physiology* showed shorter bursts of interval workouts resulted in equal increases in exercise performance when compared to longer, moderate-paced exercise sessions.

Roughly thirty to sixty minutes of daily cardio exercise is recommended for weight loss and metabolism raising; interval-training sessions are usually closer to thirty minutes in length, especially for beginners. If you want to work out for a longer period but can't sustain intervals for more than about twenty minutes, try beginning your workout with a twenty-minute relaxed pace, then end with twenty minutes of interval training. You can practice interval training with just about any form of cardiovascular exercise, including walking, running, cycling, rowing, swimming, and using an elliptical trainer. Remember, however, that interval training can be taxing; especially if you're not used to exercising intensely, you may find that you're quite a bit more tired than you expect to be after the first couple of sessions.

9 Make Exercise More Effective with Mindful Breathing

Breathing correctly while exercising can help maximize metabolic gains and improve your overall exercise performance. According to Christopher Guerriero, founder and CEO of the U.S. National Metabolic and Longevity Research Center, "The more oxygen you take in, the harder your muscles can work, the greater their growth and development, and ultimately the faster and more efficient your metabolism becomes."

Supply Cells with More Oxygen and Burn More Fat

When oxygen combines with an energy molecule (which comes from the break-down of fat and carbohydrates), your body creates energy. As bodybuilder, personal trainer, and nutritionist Tom Venuto explains, "How much fat is burned during exercise depends on the ability of the cardiovascular system to deliver enough oxygen to the cells. To speed up fat metabolism and burn off fat stores we need to supply cells with sufficient oxygen during exercise." Athletes often have more arteries and capillaries than non-athletes—up to twice as many, in fact—and these arteries and capillaries can carry more oxygen to the muscles. This translates into more fuel being burned over a shorter period of time.

The body uses oxygen to manufacture adenosine triphosphate (ATP), a chemical that provides the energy that allows you to move your muscles. When you engage in strenuous activity, you use more oxygen. To speed your metabolism, the body needs larger than normal amounts of oxygen. To keep up with this need, your heart responds by pumping faster and you start breathing harder and faster to take in more oxygen. "Your oxygen delivery system is crucial to your full metabolic development, because it is the main provider of ATP in any exercise session that lasts more than about three minutes," Christopher Guerriero explains.

Breathing Correctly Raises Metabolism

The correct way to breathe involves inhaling and exhaling through your nose, deeply and fully, so that both your chest and your abdomen move with each breath. Shallow or "chest only" breathing "results in a lower oxygen supply, which slows all bodily functions, including our metabolism," explains Dr. Tom Goode, co-founder of the

International Breath Institute, based in Texas. Conversely, breathing fully and deeply increases the amount of oxygen moving through your bloodstream, which in turn lets your body burn more calories and more fat. So, if you breathe correctly, you'll enhance your metabolism and use up more calories, even when you're not exercising.

When engaged in normal activity, you should take twelve to twenty breaths per minute; this will increase when you exercise. Establishing a regular breathing pattern is important, too. For example, inhale every five steps when you're walking or every third stroke while swimming. Establishing a pattern that gives you the most energy and endurance will help ensure optimum oxygenation and raise your metabolism.

When lifting weights, try inhaling during the let-down, or return of the weights, and exhaling while pushing or pulling the weight into position. Never hold your breath while lifting weights. A 2007 article published by the Mayo Clinic suggests that if you do, you run the risk of temporarily increasing your blood pressure to dangerous levels. Also, by increasing abdominal pressure, you may increase the likelihood of developing a hernia or hemorrhoids as a result of your exertions.

10

Use Yoga to Fan the Metabolic Flames

Fire up your metabolism with yoga. Studies show it can build strength, tone muscle, and improve circulation and thyroid function, all of which will boost your metabolism. As a 2004 article by Timothy Burgin in *New Life Journal* explains, "Metabolism consists of the chemical processes that create energy in the body that are regulated by the endocrine system, especially the thyroid. Yoga has a powerful stimulating and strengthening effect on the endocrine organs and thus can boost metabolism to burn more calories."

Stimulate Your Thyroid

If your thyroid is underactive, your entire metabolic rate may be slowed. A study reported in the March 24, 2008 issue of the *Archives of Internal Medicine* found that even slight increases in thyroid-stimulating hormone (TSH) levels within the normal range were associated with increases in body weight. Yoga poses that affect the neck area, site of the thyroid, may help to stimulate the gland. They include the hatha yoga pose, or *asana*, known as "the fish," as well as various back bends and shoulder stands, such as the bow, the bridge, the plow, and the camel. These *asanas* are often performed in a sequence to harmonize with the flow of blood through the internal organs. They act like an internal massage, stimulating the glands and organs to increase their activity and efficiency. Moving quickly between the poses can accelerate the process.

Improved Circulation Increases Energy

Jasmine Lieb, a Los Angeles-based yoga instructor at Yoga Works, whose therapeutic yoga classes have been recommended by doctors and physical therapists across the country, recommends a popular series of poses known as the Sun Salutation, emphasizing back bends "to increase circulation and energy in the entire body … Back bends are energy boosting, which rev you up."

These yoga poses may also improve circulation. "The twisting and compressing of the yoga postures massage the endocrine and abdominal organs, regulating their function, improving local circulation," explains Timothy Burgin in his 2004 article. Lieb concurs. "As circulation is increased in the thyroid, adrenal, and pituitary glands that send out endorphins and hormones throughout our bodies, balance can be restored in our metabolic system."

Standing poses, such as the warrior, and lunges offer additional metabolic benefits by building strength and toning muscle. As it takes more energy—and more calories—to maintain muscle tissue than fat tissue, incorporating strength-building postures into a yoga routine will increase caloric output. Start by holding standing poses for thirty seconds, then slowly increase the amount of time you hold a pose—you'll strengthen your muscles and improve your endurance, which ultimately will boost your metabolism.

Turn Up the Heat to Burn More Calories

Some forms of yoga can provide an intense aerobic workout. One called ashtanga yoga (or "power" yoga), involves performing a series of positions one after the other in a more rapid fashion, without resting between poses. Because you perform the movements at a high level of activity, your heart rate, breathing, and circulation are increased. Ashtanga yoga is an effective cardiovascular exercise that gets your heart rate up into the "fat-burning zone," which equates to between 65 and 80 percent of its maximum—the rate a 2002 article in the journal *Medicine & Science in Sports & Exercise* suggests is an efficient rate for burning fat (▶5).

Another intense yoga style called bikram yoga (or "hot" yoga) is performed in a hot room to help raise body temperature and burn off calories. The heat causes your body to burn more calories than it would during a typical hatha yoga session of the same length. Doing ninety minutes of bikram yoga in a room heated to 110 degrees Fahrenheit (43°C) burns between 300 and 600 calories. Advanced practitioners of bikram yoga can burn between 500 and 1,000 calories per session.

Although these more intense forms of yoga aren't suitable for everyone, hatha yoga can be done by people of all ages and body types. Most yoga poses have two or three different levels of difficulty, making them suitable for a range of experience and abilities; test your limits by trying the harder levels each time. Of course, the best way to learn to perform metabolism-boosting yoga poses effectively and safely is to join a class led by a qualified and experienced instructor. Yoga *asanas* can also provide a good complement to other forms of aerobic exercise and strength training.

Increase Your Range of Motion with Tai Chi

The Asian martial art known as tai chi (pronounced *ty chee*), or tai chi ch'uan, takes your muscles and joints on a full tour of their range of motion—and experts agree that keeping the joints moving smoothly is key to a successful workout. A 2008 study published in the journal *Clinical Sports Medicine* found that because tai chi movements require a good deal of balance and agility—as well as flexibility—they can help increase strength.

A Low-Impact Way to Burn Calories and Build Muscle

The basic principles of tai chi involve balance, coordination, and stamina achieved with relaxation, as opposed to muscle contraction and tension. Part of traditional Chinese medicine, tai chi uses movements that are slow and repetitive, giving the practitioner the ability to focus on breathing and circulation, rather than on a complex series of punches, kicks, and blocks as is the case with other martial arts, such as karate.

Tai chi moves at a gentle pace and is one of the lowest-impact exercises you can find, making it ideal for any fitness or ability level. At the same time, its practitioners can work up a sweat. When done for thirty to sixty minutes per day, tai chi will build muscle and burn calories, a proven combination for increasing metabolism.

As an example, a 150-pound (68 kg) person doing tai chi for an hour can burn 280 calories—that's about half what she would burn bicycling at a moderate pace for the same amount of time. According to a 2005 article by Bill Douglas, tai chi expert and author of *The Complete Idiot's Guide to T'ai Chi and Qigong*, tai chi provides roughly the same cardiovascular benefits as moderate-impact aerobics.

Improve Circulation and Energy

The movements of tai chi are split into two basic groups: solo movements and movements that involve pushing with the hands. Solo movements emphasize circulation, breathing, and posture; the pushing movements include defensive motions that focus on follow-through, changing your center of gravity, and various defensive positions. Throughout all tai chi exercises, practitioners work on improving their concentration and aiding the flow of life energy, known as *qi* or *chi*, through their bodies.

To increase the effects of tai chi on your metabolism, take it up a notch. Many of the movements can be done at various paces. Most novices start at a slow rate of movement, but there's no reason not to pick up the pace. While learning the proper form from a tai chi master, try to get your heart rate up to the point where you're not out of breath and are still able to talk, but carrying on a conversation becomes difficult. You can burn more calories and reap more of the benefits of a cardiovascular workout this way, while still taking advantage of tai chi's calming, balancing, strengthening, and limbering benefits.

12

Breathe Your Way to a Healthy Metabolism with Qigong

Regardless of your physical capabilities, you can raise your metabolism by taking up qigong (pronounced *chee kung*), a form of breathing, movement, and meditation that is an integral part of tai chi (▶11). Its name means, literally, "breathing exercise," "breath work," or "energy work." As with other kinds of efficient breathing (▶9), qigong increases the amount of oxygen moving through your bloodstream, which in turn raises your metabolism and enables your body to burn more calories and more fat. A 2008 paper published in the journal *Applied Psychophysiology and Biofeedback* described a study that analyzed the breathing of a group of yoga instructors. It showed that active deep-breathing exercises increased metabolic rate, compared to the quiet deep breathing of a meditative state.

Breathing Exercises Generate Energy and Increase Metabolic Rate

When your body becomes more efficient at distributing oxygen, your ability to burn more fat cells and experience greater fuel efficiency increases. Oxygen is coupled with previously stored metabolites to release energy for immediate use, as opposed to food sources of energy, which are often intended for long-term energy storage. And it's the long-term storage process that can lead to weight gain.

Qigong focuses on the energy that exists in all living things, known as *qi* or *chi*, and how it nourishes the body. It involves physical positions, special movement techniques, and breathing practices to promote mind-body balance. When performing qigong, you move from one position to another in a rhythmic, controlled, fluid motion, which builds both stamina and muscle strength. The act of attaining and holding qigong's more difficult positions can enhance muscle tone and improve circulation, two factors that will help raise metabolism.

Improve Eating Patterns to Raise Metabolism and Lose Weight

Practicing qigong can also help you lose weight by reducing your caloric intake. Jasmine Lieb, a Los Angeles-based instructor at Yoga Works, explains that "through breath and movement, more oxygen and nutrients are introduced to the blood, and toxins that interfere with metabolic balance are released. Even in a short period of time, [food] cravings and fatigue can be reduced, and with regular practice, weight loss is possible." In addition, qigong may improve digestion. Stretching, positioning, and special breathing techniques can facilitate the flow of nutrients in the body, which in turn may help you get the most from what you eat, even if you eat less.

A 1999 study published by the *Japanese Journal of Biometeorology* suggested that metabolism is elevated when someone stands or sits upright with good posture, instead of sitting, slumping, or lying down with folded limbs. Qigong not only makes you more aware of proper body positioning while eating, but its emphasis on improved posture also enables you to achieve a small improvement in your resting metabolic rate.

Qigong also affects metabolism on a secondary level. The combination of positioning, stretching, and breathing improves coordination, flexibility, and stamina. This paves the way for more efficient cardio workouts or strength-training exercises, and may help prevent injury as a result of those workouts. Although qigong will not replace vigorous exercise, it is an excellent addition to any health-maintenance routine because it helps enhance the mind-body connection.

13

Steam Off Calories in a Sauna

Sweating raises your metabolism, which means you'll burn calories and fat by basking in the heat of a sauna. According to a 2003 article published in the journal *Experimental Biology and Medicine*, clinical studies of thermal therapies (including saunas) found that just two weeks of sauna therapy improved vascular function and fat-burning in study participants.

As your environment heats up in a sauna, your body has to cope with the heat in order to keep its temperature normal. Extreme temperatures—hot or cold—can increase metabolic rate by up to 20 percent. Your heart has to work harder to boost circulation, thus improving your cardiovascular system. This extra work translates to a more intense metabolic rate, which may

result in burning up to 300 calories during a thirty-minute sauna session. And you may continue to burn more calories for up to three hours afterward.

Losing Body Weight as Sweat

We humans each have more than 2.5 million sweat glands, covering almost every part of the body. When stimulated by heat from exercise, temperature, or stress, these tiny tubes full of cells produce sweat by secreting a water and electrolyte mixture. In an attempt to maintain normal body temperatures, you can sweat between one and three liters (33 and 100 fl oz) an hour when exposed to thermal stressors.

Sitting in a sauna for thirty minutes produces approximately the same amount of perspiration as a six-mile

(10 km) run. As Susan Smith Jones, Ph.D., explains, you can literally sweat off one pound (0.45 kg) simply by sitting in a sauna for half an hour—approximately as much as you would rowing for the same amount of time.

Saunas can encourage weight loss in another way, too, because body fat becomes water-soluble at approximately 109 degrees Fahrenheit (43°C). Therefore a sauna experience allows the body to sweat out excess fat. This can also allow the body to start eliminating fat-stored (lipophilic) toxins, which some people find aids weight loss.

Varieties of Sauna, Old and New

Saunas have been used in Finland for more than two thousand years, and for centuries Native Americans have utilized sweat lodges, which are similar in many ways. Modern saunas come in two basic varieties: wet and dry. Wet saunas (sometimes called steam rooms) typically use fire and volcanic rocks, or an electronic equivalent, to generate a heated environment. The rocks retain the heat and keep it from dissipating too quickly. Water splashed over the rocks creates steam that moistens the air. Wet saunas can reach temperatures of 100 to 115 degrees Fahrenheit (38 to 46°C).

Dry saunas also use superheated rocks, but the temperatures are much higher, up to 200 degrees Fahrenheit (93°C), because the air is devoid of most moisture and thus has a higher heat capacity. A newer type of sauna, the infrared sauna, uses infrared radiation to create the heated environment; temperatures in these saunas tend to be between 120 and 140 degrees Fahrenheit (49 to 60°C).

Increased Heart Rate and Circulation Boost Metabolism

Steam bathing has a stimulating effect on the cardiovascular system, too. A fifteen- to twenty-minute session can increase your pulse rate from seventy-five beats per minute to between 100 and 150 beats per minute. Although blood circulation is increased—which also boosts metabolism—blood pressure isn't elevated because the heat expands the blood vessels in the skin, enabling them to handle the increased blood flow.

All this may sound like an easy, exercise-free way to boost your metabolism and lose weight fast. The downside, however, is that the calorie shedding is only temporary; the weight loss incurred during the sauna is likely to be water-loss from perspiration. And remember that sitting in a sauna does not work any muscles other than your heart. Nonetheless, a sauna is another tool to help raise your metabolic rate—a pleasant adjunct to cardio exercise and strength training. Use a sauna not as an exercise replacement, but rather as a supplement to your current regime.

Note that senior citizens, babies, pregnant women, or people with heart conditions should consult their doctor before stepping into a sauna or steam bath. As with all exercise and health routines, listen to your body. If you feel nauseated, dizzy, or lightheaded, exit the sauna immediately.

14 Get a Daily Dose of Exercise

The more exercise you can incorporate into your daily routine, the more efficient and speedy your metabolism will be, and the better you'll feel. Research, such as a 2006 study published in the *Journal of Physiology*, supports the idea that short, intense exercise can be just as beneficial as longer workouts of lower intensity. The key is to exercise regularly, whenever and wherever you can. You may not have an hour or two a day to put in at the gym, but a study by Professor James Timmons and researchers at Heriot-Watt University in Edinburgh, Scotland (published in the journal *BMC Endocrine Disorders*), found that even brief exercise sessions—as little as four to six sprints of thirty seconds each on an exercise bike, every other day—had a marked impact on how the body metabolizes sugar.

Incorporate Exercise in Your Lifestyle

Exercise was an essential, natural, and integral part of our ancestors' lives. With the development of technologies that "make life easier," our whole society has become less active and more obese. Today, exercise is generally seen as an adjunct to our other commitments, something we do because we know we should. You can change that by incorporating exercise into your daily lifestyle. Walk or bike to town whenever possible, instead of driving. Go out and play sports instead of watching professional athletes on television.

It's the cumulative effect of daily exercise that makes metabolism more efficient, so make exercise a habit. Daily exercise needn't be as formalized as kick-boxing to music or using an elliptical trainer for thirty minutes. In many southern European countries, many people take a walk after meals—sometimes for hours—to enhance metabolism and digestion. Moving food and nutrients through the body efficiently is a major reason to exercise. Metabolism, in part, is about how your body translates the nutrients you eat into energy for your body, so walking after meals can result in a direct boost to your metabolism. A 1991 study in the *Journal of Applied Physiology* found that the metabolic effects of exercise can help you metabolize your food more efficiently—and burn off calories—for as much as four hours after eating. And with a daily dose of exercise, your metabolism also becomes more efficient because of the increased amount of muscle tissue in the body.

15

Burn Calories on the Sly with Stealth Exercise

Even if you can't factor a longer walk or workout into your daily schedule, you can improve your metabolism with "stealth exercise"—squeezing exercise into your usual routine. Your metabolism (the rate at which you burn calories) increases every time you expend energy, so every little bit counts (▶14).

Exercise Whenever You Can for Long-term Gains

Stealth exercise can be done in a number of different ways throughout the day, without requiring a large chunk of time, workout gear, or a fitness-club membership. It is true that intense activity stretched out over time raises metabolism more efficiently because sustained exercise burns fat. And yes, it's ideal to engage in at least half an hour of daily exercise. However, small nuggets

of exercise here and there burn calories too, and those calories add up, especially if you're mindful of the fact that these activities can improve your health.

The key to stealth exercise is to get up and get moving. All exercise adds up, so try adding in some physical activity whenever you have a few free minutes. Remember that a pound (0.45 kg) is roughly equivalent to 3,500 calories. If you consistently can burn an extra 500 calories per day exercising—without replacing them with extra food of course—over the span of a year you could lose up to fifty pounds (23 kg).

Walk More, Weigh Less

Walking is one of the simplest forms of exercise, and studies such as one published in 1999 in the *New England*

Journal of Medicine show that walking can be very effective in encouraging fitness and preventing disease. Results presented at the 2008 meeting of the American College of Sports Medicine revealed that walking is a safe and effective exercise for even very obese people, so it is a great way to begin to introduce exercise into your routine.

A good way to incorporate walking into your daily routine is to steer clear of that coveted parking spot near the mall or supermarket entrance. Instead, park in the rear of the lot, as far from your destination as you can get. For a woman weighing 150 pounds (68 kg), five minutes of walking at a moderate pace can burn twenty-five calories. That may not sound like much, but if you run errands at two stores and park far away

stairs is a powerfully efficient aerobic exercise and burns more calories than you would in a moderately paced aerobics class. Increase the pace and duration to maximize your metabolic benefit (providing you are fit for the activity from a medical point of view).

Dont Just Watch, Join In!

If you're doing errands that don't require hauling anything large or awkward, bicycling is a good alternative to driving. The number of calories burned depends on your pace and the terrain, but riding on city streets for half an hour burns an average of 250 calories. Working in the yard is another way to boost metabolism. Actively weeding, watering, raking, and planting can use up as much as 350 calories an hour, which is approaching the amount you might burn in a gym workout.

Join your children in their myriad physical activities. Don't just watch them jumping rope or chasing each other around a park—take part, and burn as much as an extra 600 calories an hour. Play kickball to work off up to 500 calories an hour, or toss a Frisbee with them to lose up to 250 calories an hour. Instead of standing on the sidelines, play mini-golf—getting in on this kind of action could burn up to 200 calories an hour.

from the entrance both times, you'll have burned one hundred calories by the end of your trip—and that's not counting the walking you do *inside* the stores as part of doing your errands.

Walking up stairs for just five minutes can burn up to fifty calories. If you only intend on going to the second floor, take a few more minutes to walk up and down a couple extra flights. Climbing

PART II

You Are What You Eat: Nourish Your
Metabolism with Good Nutrition

16

Eat Complex Carbohydrates to Lose Weight

Get more bang for your calorie buck by eating complex carbohydrates rather than refined ones. The Asian Network for Scientific Information, in 2002, published research showing that consumption of certain complex carbohydrates was associated with lower body weight, reduced blood cholesterol, and reduced blood glucose. In addition, the Linus Pauling Institute, a micronutrient research center at Oregon State University, points out that eating whole grains aids digestion, which can contribute to weight loss.

The glycemic index provides another way to look at carbohydrates. The glycemic index rates carbohydrates in terms of how much they raise blood sugar levels during the several hours after you consume them. The GI uses a 1-to-100 scale: higher-numbered foods will convert to glucose (and raise blood sugar levels) quicker than lower-numbered foods. A 2004 study published in the *Journal of the American Medical Association* showed that reducing your glycemic load, by consistently eating foods that are low on the glycemic index, can help regulate metabolism and promote weight loss.

Slower Sugar Metabolism Means Less Stored Fat

Carbohydrates are composed of sugars; your metabolism converts these sugars to glucose, which your body uses for energy. A complex carbohydrate is a type of carbohydrate that contains sugars that are bonded together. They therefore take longer to digest, meaning in turn that the sugars are absorbed more slowly by your body. Slower breakdown results in smaller, steadier blood sugar levels—and your body is less likely to attempt to rapidly lower its blood sugar levels by storing the excesses as fat. Simple carbohydrates, on the other hand, are digested quickly and, when eaten in large amounts, are more likely to be converted to fat.

"One of the biggest benefits of complex carbohydrates is that they … will be broken down very slowly by the body, resulting in constant fueling of energy," explains power lifter Marko Dimitrov. "Also the amount of calories spent for digesting complex carbohydrates is certainly bigger than the one for simple carbohydrates, which is excellent because you'll spend additional calories which you don't need."

The American Heart Association recommends that you get approximately 55 to 60 percent of your energy intake from carbohydrates—especially complex carbs. If you're engaged in strenuous activity, that could increase to as much as 70 percent, according the Canadian Dietetic Association. The U.S. Department of Health and Human Services suggests eating complex carbohydrates three times daily.

Whole Grains Raise Metabolism

Examples of complex carbohydrates include whole-grain and brown breads, brown rice, whole-grain pastas, potatoes and root vegetables, peas and lentils, and oats. These foods provide more energy than simple carbohydrates and can help reduce the amount of excess food that is converted to fat.

Increased intake of whole grains can also help improve your overall health, giving you a strong baseline to raise your metabolism. Simple, processed carbohydrates are found in candy, sugary sodas, cakes, pies, and cookies, and regular granulated table sugar.

Look for "whole-grain" breads and pastas. All grains are composed of three elements: bran, germ, and endosperm. Processed and refined grains consist of only the endosperm, whereas whole grains retain the entire grain kernel. Bran and germ contribute much of the protein found in grains, and also provide other valuable nutrients.

As explained on page 12, metabolism can be seen as divided into two basic processes: anabolism and catabolism.

Anabolism is the process that creates the matter the body requires; catabolism is the process whereby the body breaks down fuel sources to supply energy. If you eat a good range of complex carbohydrates, catabolism breaks down those carbohydrates steadily and allows your body to use them more efficiently. When you eat too many simple carbohydrates, anabolism kicks in to create new fat cells from that excess.

Calorie for calorie, nutrient-dense, fiber-rich foods provide more satisfaction than high-fat, high-sugar foods—so you don't have to eat as much to feel full. Simple unprocessed carbohydrates (fruits such as apples, oranges, stone fruits, and berries) contain low amounts of simple sugar, and can also be part of a healthy, metabolism-boosting diet.

Beef Up Your Metabolism with Protein

Feast regularly on a good steak to beef up your metabolism. Protein requires more energy to digest than carbohydrates or fats. The extra energy your body must expend to digest protein results in a temporarily increased metabolic rate. This effect is sometimes referred to as "meal-induced thermogenesis." In general, as noted in the 2002 edition of the standard reference work *Introduction to Human Nutrition*, about 10 percent of the calories in a given meal are used up by the process of digestion; however, when eating protein it's possible to burn as much as 25 percent of the calories consumed through digestion alone, as recorded in the 1986 book *Exercise Physiology: Energy, Nutrition, and Human Performance* and a 2004 review in the journal *Nutrition & Metabolism*.

A secondary effect of eating protein can contribute to weight loss, too. Because it takes longer to digest, protein keeps you feeling full for longer, as noted in a 1999 study in the *European Journal of Clinical Nutrition*. The longer you feel full, the less likely you are to overeat—and reducing your total number of daily calories is the key to both losing weight and regulating metabolism.

Eat Protein to Build Muscle

Proteins are molecules that are made up of a chain of amino acids. They come in two basic types, simple proteins and conjugated proteins, and each performs a specific function. Proteins are necessary for regulating and maintaining cells and tissue. So essential are they to the body's functioning that they're often called the "building blocks" of the body.

Studies, such as a 2002 article published in the journal *Medicine & Science in Sports & Exercise*, show that protein intake is essential to helping muscles and tissues recover and rebuild after a workout. For this reason, some experts recommend a post-workout snack of both carbohydrates (to replenish glycogen stores) and lots of protein (to help tissue recovery).

In the United States, the government-issued recommended daily allowance (RDA) is fifty grams (almost 2 oz) of protein per day. If you're trying to maintain the shape you're in, about 0.8 grams per kilogram (2.2 lb) of body weight per day is usually recommended. If you want to add muscle mass, you'll need to consume 1.2 to 1.8 grams per kilogram of body weight.

Good Sources of Protein

There are many creative ways to get enough protein into your diet—whether you're a vegan, a carnivore, or somewhere in between. Meat and fish are some of the best sources of complete protein. Fish, chicken, beef, pork, and shrimp all have around twenty-two grams of protein per three-ounce (85 g) serving—about the size of a deck of cards. Yogurt contains about eleven grams of protein per cup, milk has eight grams per cup, and most cheese has approximately seven grams of protein per ounce (28 g). Eggs contain seven grams of protein each. When eating meat as a protein source, make sure to choose lean varieties. Although bacon has two to three grams of protein per slice, it also contains two to three grams of fat—per slice!

57

Vegetarians also have plenty of protein-laden options, too. A half-cup serving of tofu contains about fifteen grams of protein. Lentils are also a good source of protein, with nine grams in a half-cup. A half-cup of walnuts has about nine grams of protein, and a half-cup of sunflower seeds has about five grams. Peanut butter contains around four grams of protein per tablespoon, chickpeas (which are found in hummus) contain seven grams per third-cup, and grains such as kasha and bulgur wheat have around three grams per half-cup. Many breakfast cereals are also infused with extra protein. Supplies of protein can also be found in broccoli, spinach, and kale; cashews; beans, peas, and soy; and whole-grain breads and pastas, brown rice, oats, muesli, and barley.

When protein-laden foods are eaten and digested, they break down into amino acids (▶18). The body requires twenty different kinds of amino acids, twelve of which (called non-essential amino acids) can be produced in the body itself. The remaining eight, known as essential amino acids, must be consumed in order for the body to function normally. Complete proteins (found in meat, eggs, dairy products, and other animal sources) contain all of the essential amino acids. Incomplete proteins, on the other hand, contain only some of them. Therefore, you'll need to combine at least two incomplete proteins in a meal in order to create the equivalent of a complete protein.

18

Build Muscle and Burn Fat with Amino Acids

Eat protein-rich food after a workout to strengthen and repair your muscles. Muscle tissue cannot regenerate and grow without the critical support of amino acids, especially those found in protein. Studies, such as one published in a 2006 article in the *Journal of Nutrition*, show that branched-chain amino acids, or ones with a particular type of branched-chain structure, are found in greater quantities in muscle tissue, and they may actually stimulate the creation of muscle protein.

When you eat a high-protein food (▶17), digestion breaks the protein down into its component amino acids. As they enter the bloodstream, these amino acids are sent to work on tasks of great importance, such as rebuilding muscle and repairing tissue damage.

These are vital operations that allow muscles to repair and grow stronger following a weight-training session.

Amino Acids Help Reduce Fat and Weight

As your percentage of lean muscle mass grows, so does your fat-burning potential. Studies, such as the above-cited one published in the *Journal of Nutrition*, suggest that branched-chain amino acids including L-valine, L-leucine, and L-isoleucine may be more effective at building muscle than other types of amino acids. Bodybuilders may supplement with these specific acids, but the casual meta-booster need look no further than her supermarket: Most protein-laden foods contain branched-chain amino acids, particularly dairy products and red meat.

Amino acids can help in other ways to raise your metabolism and lose weight. The amino acid L-tryptophan might diminish cravings for sweets and carbohydrates, says Michael Rosenbaum, M.D. Additionally, the amino acid called L-phenylalanine may decrease appetite by increasing the body's production of norepinephrine, working much like an amphetamine would.

According to an 1989 article in the magazine *Better Nutrition*, L-carnitine, L-tryptophan, and L-phenylalanine may help in the fight against obesity by burning fat more efficiently and suppressing appetite. Taurine, an amino acid found in large quantities in the brain, is touted for its energy-boosting abilities. However, despite taurine's presence in many energy drinks, there is little solid proof that taurine supplements can cross the blood-brain barrier to increase overall energy.

The Building Blocks of Proteins

Amino acids are organic, meaning they are found in living beings. Amino acids contain the basic elements of carbon, nitrogen, hydrogen, and occasionally sulfur. The molecules of these acids are bonded together in chains known as peptides; groups of amino acids bonded together are called polypeptides. Put the polypeptides together and what do you have? You guessed it: protein. Protein molecules are simply a combination of amino acids and peptides.

Two classes of amino acids are needed for human development: essential and non-essential. While both are necessary for a well-rounded, healthy diet, the essential amino acids must be obtained from foods, drinks, or supplements—they aren't produced naturally in the body.

The specific amino acids that are essential include L-phenylalanine, L-leucine, L-methionine, L-lysine, L-isoleucine, L-threonine, L-valine, L-tryptophan, L-histidine, and L-arginine. The non-essential amino acids include L-taurine, L-tyrosine, L-alanine, L-arginine, L-asparagine, aspartic acid, L-cysteine, L-cystine, glutamic acid, L-glutamine, L-glycine, L-omathine, L-proline, and L-serine. Many of these names may well look familiar to you. That's probably because some of them are included in popular energy drinks— and now you know why.

Unless you're an elite athlete or are working with a licensed nutritionist, it's probably not a good idea to supplement with targeted amino acids. There may be health consequences to deliberately inducing an amino-acid deficiency. Focus instead on eating a diet that consists of a wide range of healthy foods.

19

Spike Your Metabolism with Spicy Foods

If you love hot Indian, African, Mexican, or Szechuan cuisine, you'll be glad to learn that eating spicy food can actually raise metabolism and burn fat. Studies, beginning with a 1986 analysis published in the journal *Human Nutrition: Clinical Nutrition*, have shown that you may burn more calories when you eat spicy foods such as chili, hot pepper, or mustard with a meal.

A wide range of foods is available in spicy, spicier, and ridiculously hot varieties. Some of the hottest foods are the capsicums, the family from which the cayenne pepper derives. Peppers of this sort contain an ingredient called capsaicin. The chemical irritant in capsaicin burns the mouth and the skin (which is why many chefs wear gloves when preparing especially hot peppers), but it is also thought to help elevate metabolism. Chili peppers, such as the jalapeno, are part of the same family.

Hot Food Burns Fat

When you consume food, your metabolic rate rises slightly. This is called the "thermic effect" or "thermogenesis." Spicy foods, according to a 2006 article in *Physiology & Behavior*, are said to increase thermogenesis (as compared to non-spicy foods), and give off heat as they increase the rate at which the body burns fat tissue. The fat-burning effect can last up to a few hours after eating a spicy food, meaning you may be burning additional fat stores during this time.

Recently, clinical studies have attempted to measure the effectiveness of spicy foods in raising metabolism and burning fat. A 2007 study published in the *International Journal of Obesity* found that participants in a double-blind weight-loss trial who took a supplement containing spicy extracts, such as tyrosine, capsaicin, and catechines, lost more weight and had a greater thermogenic effect than those who took placebos.

Some spicy foods fool the brain into thinking that the body is being exposed to another hot stimulus, hence the sweating. Some people are more susceptible to this than others. Spicy foods can stimulate body temperature and circulation, both of which can induce sweating and raise metabolism (▶13).

Spicy Foods Induce Satiety

Eating spicy foods may naturally limit food intake as well. A 2006 article

published in the journal *Physiology & Behavior* suggested that eating spicy foods can increase your feeling of satiety and thus prevent overeating. That's because when spicy food leaves a burning sensation in the mouth, many people choose to wait a while until the fire subsides before eating more. By that time, the stomach has had a chance to communicate a feeling of fullness to the brain, which decreases hunger signals.

Some people, though, should avoid spicy foods. If you suffer from chronic heartburn, spicy foods tend to exacerbate that condition. People who have ulcers or other intestinal problems might want to stay away from spicy foods, too.

20

Add Zing to Your Metabolism with Ginger

Here's more good news for metabolism-boosters who love Asian, African, and Indian cooking. Fresh ginger *(Zingiber officinale)*, the strong-tasting plant that is a component of many spicy dishes, not only adds zing to meat and vegetables, it can also raise your metabolism.

Ginger contains active ingredients that are structurally similar to capsaicin, a spicy compound believed to help elevate metabolism (▶19). For this reason, ginger is said to be a "thermogenic" food, meaning the body has to do some extra heat-producing work to digest it. Studies, such as one published in a 2006 article in the journal *Obesity*, have shown that drinks made from thermogenic ingredients lead to increased energy expenditure for up to twenty-four hours following ingestion.

Fat Oxidation Can Help Prevent Obesity

Research shows that ginger also aids your ability to metabolize both fat and protein. According to an article published in *Physiology & Behavior* in 2006, ginger's thermogenic properties can significantly impact fat oxidation and help prevent obesity. Additionally, the article suggests, eating ginger or drinking it in a tea can enhance your energy balance.

Although few Western scientific studies exist to provide clinical evidence of ginger's benefits to the metabolism, this powerful rhizome has been used in Asia in this capacity since ancient times. Ayurvedic (ancient Indian) medicine recommends ingesting ginger to regulate metabolism and improve digestion.

Traditional Indian medicine prizes ginger so highly it refers to the plant as a "universal remedy."

A Tonic for Your Metabolism

Adding ginger to a healthful meal can help you obtain its metabolism-boosting benefits. You might also enjoy ginger beers or ginger ale, which have long been considered tonics, used to settle the stomach and aid digestion, or a wide range of ginger-infused teas.

Ginger is not, however, a medication and does not have a recommended daily allowance (RDA) value. According to the Mayo Clinic, a safe dosage for health maintenance is around one gram per day, and no more than four grams per day. Before integrating new supplements into your diet, consult your physician.

21 Eat Fat to Stay Slim

If you think all fat is a no-no, think again. Omega-3 is a type of unsaturated fatty acid, found in a variety of foods such as fish and soy products, that is actually good for you and may help you lose weight. A 2007 study published in the *International Journal of Obesity* found that young, overweight men who ate seafood or added fish-oil supplements—which contain omega-3s—to their diets, lost more weight than those who didn't ingest the supplements or seafood.

Omega-3s Decrease Appetite

In addition, omega-3s can help you lose weight by diminishing your appetite. A 2008 study published in the journal *Appetite* showed that volunteers in a weight-loss study who supplemented their diets with omega-3s felt more satisfied and less hungry after eating than people who didn't supplement. Some experts believe that omega-3 fatty acids increase the body's secretion of a hormone called leptin (▶**48**). Leptin is a protein hormone that works to regulate satiety and appetite, with possible effects on body temperature and energy levels as well. Animal studies, as reviewed in a 1998 paper published in the journal *Nature*, have shown that supplementation with leptin can lead to a decrease in appetite and body weight; however it's not certain yet whether these studies apply directly to overweight humans.

Omega-3s may aid metabolism boosting and weight loss in other ways, too. For instance, people who eat a diet that contains sufficient omega-3 fatty acids tend to have fewer problems with blood sugar and high cholesterol, two conditions that can often derail efforts to lose weight.

Omega-3s are called "essential fatty acids," meaning that the human body requires them for optimal functioning, but they cannot be generated in the body. The difference between unsaturated and saturated fatty acids lies in the chemical makeup of their molecules. Unsaturated fatty acids have double bonds between carbon atoms, whereas saturated fatty acids have a hydrogen atom between all carbon atoms. The categories of omega-3 fatty acids we're interested in here are ALA (linolenic acid), DHA (docosahexaeonic acid), and EPA (eicosapentaenoic acid). ALA is converted to EPA and DHA during digestion.

Ways to Increase Omega-3 Fatty Acids in Your Diet

Because omega-3 fatty acids are available in a wide variety of foods, they can be easily incorporated into your diet. Several categories of foods are particularly high in omega-3s. Flax seeds are one of the best sources of omega-3 fatty acids, containing about 1.75 grams per tablespoon. Walnuts are also quite high, with about 4.5 grams per half-cup, as are pumpkin seeds (four grams per quarter-cup).

Soy products are another good source; soybeans provide around one gram per cup. Some fruits and vegetables also contain omega-3 fatty acids, including strawberries (0.1 grams per cup), spinach (0.3 grams per cup), green beans (0.1 grams per cup), kale (0.15 grams per cup), and broccoli (0.2 grams per cup).

Many types of fish—especially oily species—are high in omega-3 fatty acids. Salmon and mackerel each provide two grams per four-ounce (113 g) serving. Herring and tuna each have about one gram per four-ounce (113 g) serving. Halibut offers 0.6 grams and shrimp 0.4 grams per four-ounce (113 g) serving. Cod liver oil, though not as tasty as smoked salmon, is another good source of omega-3s.

How much omega-3 fatty acid should you try to consume every day? There are no recommended daily allowances (RDAs) for omega-3s; however, nutritionists generally recommend that someone eating a 2,000-calorie-a-day diet should consume approximately one gram of omega-3s per day.

Always consult a knowledgeable nutritionist or physician if you plan to add supplements to your diet.

Melt Weight Away with a Grapefruit a Day

Eat a grapefruit a day and the fat melts away! Take grapefruit extract and become thinner overnight! The renowned "Grapefruit Diet," which first appeared in the 1970s, seems to make unrealistic promises. But scientific evidence suggests that eating grapefruit *can* contribute to both weight loss and raising metabolism.

Eat Grapefruit to Metabolize Glucose

Naturally low in calories (half a grapefruit only contains about 40 calories), grapefruit has been documented to reduce insulin levels in the body. This reduction affects glucose metabolism and can aid weight loss.

A 2006 study published in the *Journal of Medicinal Food* measured insulin levels after glucose ingestion in people who ate grapefruit or grapefruit extracts versus those who ingested a placebo. The results indicated that the grapefruit eaters had markedly lower post-glucose insulin levels. Not surprisingly, the grapefruit group also lost more weight during the twelve-week study. Reduced insulin levels may also have prompted participants who ate half a grapefruit with meals for twelve weeks to lose an average of three pounds (1.4 kg), according to a 2006 study funded by the Florida Department of Citrus.

In the grapefruit world, color matters. Pink-and-red grapefruit is higher in antioxidants (▶**55**), which prevent cell damage, and much more effective than the white variety at lowering triglycerides (a type of fat found in the blood), according to a study carried out in Israel and reported in *The 150 Healthiest Foods on Earth,* by nutritionist Jonny Bowden, Ph.D.

Fill Up with High-Fiber Fruits

Grapefruit is also high in pectin, a type of soluble fiber that may lower levels of blood cholesterol. The two major types of dietary fiber are soluble and insoluble (▶**28**). Soluble fiber attaches to the fatty acids in the stomach and slows the rate at which sugar is absorbed through the digestive system. Soluble fiber may help lower both total and LDL cholesterol.

High-fiber foods can take longer to chew, giving your body time to send signals to the brain telling it you're full. Additionally, they can take longer to digest, so you feel full longer.

Are oranges, lemons, and other citrus fruits just as healthy as grapefruit? In terms of fiber content, grapefruit does not have distinct advantages over its citrus relatives. Oranges and lemons are both high in fiber, but grapefruit contains greater amounts of bioflavonoids. While grapefruit certainly is an asset for most dieters, a 2008 paper published in the *American Journal of Clinical Nutrition* warns of potentially dangerous effects of grapefruit on a number of important prescription medications.

Other citrus fruits are not known to have the same adverse interactions with medications, so for certain individuals oranges, limes, and lemons may prove a good substitute. But always consult a physician before starting any new food regime, especially if you're taking any prescription medications.

23

Eat Foods High in Calcium and Magnesium to Lose Weight

You know that milk helps build strong bones, but did you know it can also help you lose weight? People who eat diets high in calcium tend to weigh less—and lose weight more easily—than people with a lower calcium intake. A 2005 study of people trying to lose weight, published in the journal *Obesity Research*, indicated that test subjects who consumed at least 1,200 milligrams of dairy calcium per day burned substantially more fat from around their waists than people who consumed the same number of calories but less calcium. These researchers also found that calcium had a protective effect against the loss of lean muscle mass, a major driver of metabolism. A 2001 article published in the *Journal of the Federation of American Societies for Experimental Biology* demonstrated that calcium helped prevent fat storage and decreased the size of existing fat cells in mice.

An Effective Tag Team

Magnesium, which works in conjunction with calcium, is equally important to your metabolism. At its most fundamental level, magnesium is essential for metabolism boosters because it helps your muscles to contract and relax properly. Magnesium is vital in the production of ATP (adenosine triphosphate), the most basic form of energy storage in the body. It also plays an integral role in the breakdown of both fats and carbohydrates into energy.

Calcium and magnesium are one of the body's primary "tag teams," working synergistically to promote healthy bones and vital organs. Calcium is a chemical element required by all living organisms, and is one of the metals (yes, it's a metal) found in the highest concentration in many living things. It helps regulate your hormones and aids blood circulation, muscle contractions, and other vital functions. Calcium drives heart muscle contractions, and magnesium controls the relaxation in between heartbeats. Magnesium, in addition to contributing to the strength of your bones, helps nerves and muscles work properly. It also promotes proper heart function, keeps the immune system and blood sugar levels healthy, and helps maintain consistent blood pressure. Low supplies of calcium and magnesium in the body can have devastating effects on your cardiovascular, skeletal, and muscular health.

What are the Best Sources of Calcium and Magnesium?

Humans require two to three times as much calcium as magnesium, and these two minerals must remain in that ratio in order for the body to function optimally. The recommended daily allowance of calcium for an adult is 1,000 milligrams. Fortunately, many foods—most of them dairy products—are high in calcium.

Milk, yogurt, and cheese are some of the best sources. Milk has around 300 milligrams of calcium per cup, and cheese averages 200 milligrams per ounce (28 g). Vegans can get calcium from green vegetables such as kale, broccoli, and spinach, as well as from fortified cereals and juices. Tofu, certain types of fish, and enriched bread are additional sources. Calcium is a vital nutrient, so supplements are generally recommended if someone is unable to take in enough calcium from food alone.

An adult female requires 320 milligrams of magnesium per day, an adult male 420 milligrams. Magnesium can be found in certain types of fish; halibut, for example, has about 90 milligrams of magnesium per three-ounce (85 g) serving. Nuts (almonds and cashews are particularly good sources), whole grains, and vegetables (potatoes, spinach and other leafy greens, tofu, lentils, and peas) are also high in magnesium.

If supplements are recommended by a physician, make sure to read labels carefully. Calcium that is combined with vitamin D can be more effective than calcium alone, because vitamin D aids in the absorption of calcium. Chelated magnesium (a type of magnesium bound to amino acids) may be easier for the body to absorb and use. A combined calcium and magnesium supplement helps ensure that you'll get these two vital nutrients in the proper relative amounts. However, if your diet already contains adequate amounts of foods with magnesium, it may not be necessary to use a supplement. Excessive amounts of magnesium can produce undesirable results including muscle weakness and low blood pressure.

24

Burn More Calories than You Eat with Negative-Calorie Foods

Wouldn't it be nice to eat all you want and still lose weight? In a weight-loss study conducted at the University of California at San Francisco, Dean Ornish, M.D., helped a group of people lose an average of twenty pounds (9 kg) each. They did it by eating a vegetarian diet, without limiting their calories in other ways or undergoing any type of special exercise regimen.

Lose Weight Eating Hard-to-Digest Foods

A calorie is a unit of heat. In the language of food, calories are food energy. When digested, food gives the body energy. All foods contain calories, but a "negative-calorie food" is a food that requires more energy (calories) to digest than it supplies. For example, raw celery contains ten calories per stalk. But you burn twenty calories by chewing and digesting that stalk of celery, so eating it actually sets you ahead by ten calories. Obviously, eating such foods can help you lose weight.

Eat Veggies to Create a Calorie Deficit

Not surprisingly, most of the foods that fall into the "negative-calorie" category are vegetables and fruits. Unprocessed fruits and vegetables with pulp, skin, and fiber take the body longer to break down than processed, sugary foods, which are quickly digested and absorbed.

Asparagus, broccoli, lettuce, spinach, zucchini, and green beans, for instance, are low in calories and also use lots of energy to digest. Asparagus contains a carbohydrate (inulin) that isn't digested until it reaches the large intestine.

Several fruits, including apples, berries, grapefruit, melons, and pineapple, fall into similar categories.

Is it safe to survive solely on negative-calorie foods? The answer is indisputably no. People need protein, and some fat, for their bodies to function optimally. Inadequate intake of proteins can be especially disastrous for anyone trying to raise metabolism, as lean protein is one of the best building blocks for gaining muscle tissue (▶17)—and more muscle can translate into a higher resting metabolism.

Note that negative-calorie foods are not the same as "empty-calorie" foods, such as candy, cookies, and doughnuts, which provide virtually no nutritional value and contain high levels of fat and sugar.

25

Cut Calories Gradually to Raise Metabolism

Slow and steady does it when you're losing weight. Obviously, lowering the number of calories you consume is a logical way to shed unwanted fat and put yourself on the path to a speedier metabolism. But beware of suddenly dropping your daily caloric intake by a large percentage. Reducing your daily number of calories should be a gradual process.

One pound (0.45 kg) equates to approximately 3,500 calories. If your goal is to lose one pound (0.45 kg) per week, you'll need to cut about 250 to 500 calories per day, depending on how much you exercise. Start small. Skip dessert, saving yourself 200 calories a day, or take the cheese off your burger at lunch to save 100 calories. Gradually reduce the number of calories you eat until you reach your target number. The healthiest way to achieve this is to decrease your intake by 250 calories and increase calorie expenditure through exercise by 250.

Starvation Diets Will Only Slow Metabolism

Depriving your body of sufficient fuel can result in your metabolism actually slowing down, not speeding up. Suppose your body is accustomed to running on 2,500 calories a day. If you suddenly reduce that amount to 1,000 calories a day, your body has to make adjustments in order for you to subsist on a smaller amount of fuel. In other words, your metabolism goes into starvation mode, slowing down to maximize every nutrient it can get from what little food it's being supplied.

In the absence of incoming fuel, proteins in muscle tissue may be converted to glucose in order to supply the body with energy. Decreased muscle mass leads to burning fewer calories when the body is at rest. Because muscle is metabolically active tissue, less muscle means a slower metabolism.

What happens when you give up on the starvation plan and start to eat more normally again? You guessed it. Whatever weight you lost will likely come back, and then some, because your metabolism has slowed down to accommodate the low-calorie intake.

Maximize Fat-Burning While Cutting Calories

How can you determine the number of calories you should consume daily in order to lose weight and maximize fat burning? The exact number of calories each person needs is determined by gender, height, age, weight, and daily exercise. To maximize fat burning while cutting calories, use the formula on page 13 to determine the right number of calories for your specific body type. Figure out your BMR (basal metabolic rate), then add on the number of calories you expend every day through exercise, and then, to play it safe, add a "nudge value" of about 200 calories per day. Subtract approximately 500 calories or, if you wish to lose less than a pound (0.45 kg) per week, find the total number of calories you'd like to lose per week, and divide by seven to get your daily total. Never let your calorie intake drop below 1,200, however, without consulting a physician.

Let's use as an example a thirty-five-year-old woman who is five feet four inches (163 cm) tall and weighs 150 pounds (68 kg); she has a BMR of about 1,440 calories per day. If she works at a desk job and does little else, her daily caloric requirement is about 1,650 per day to sustain her body weight. However, if she exercises moderately for an hour every day, she'll need about 2,200 daily calories. She would need to reduce the number of calories she consumes to 1,700 per day if she wanted to lose weight while maintaining her same level of activity.

Taking this slow and methodical approach will help ensure that your metabolism does not sink, but rather increases as you lose weight and start exercising on a consistent schedule.

Remember calorie counting is not an exact science and not everybody loses weight in the same way, so do your best to make healthy choices and don't obsess over every calorie.

26

Sleep Well and Eat Early to Burn Fat

Starting the day with a healthy breakfast is one of the best ways to jump-start your metabolism and commence a day of fat- and calorie-burning fury. After a night of resting and sleeping (and, for most of us, not eating), your metabolism is running at a conservative pace. Metabolic rate declines during sleep because the body is at rest, and does not require as much energy as it does when you are awake and moving around.

A 2006 article published in the *British Journal of Nutrition* showed that sleeping metabolic rate varied from between 82 percent and 93 percent of BMR, depending on age group (children had the lowest sleeping metabolic rate). Eating a good, healthy breakfast early in the day shifts your metabolism into gear after a period of sleep.

Skipping Breakfast Leads to Unhealthy Snacking

Omitting breakfast from your daily plan is actually one of the worst things you can do for your metabolism. Studies show that people who skip breakfast tend to consume *more* calories during the day than people who eat a healthy breakfast. Many people think they don't have time for breakfast, but a nutritious, light meal can take mere minutes to prepare. If you aren't hungry when you first wake up, try to stimulate your morning appetite by exercising, and limit those late-night snacks and heavy, evening meals.

If you skip breakfast, you're likely to start feeling hungry, tired, and lethargic by mid-morning. At that point, you're much more apt to eat what's available

(doughnuts, muffins) rather than making a sensible decision to eat what's good for you (oatmeal, fruit). When you're hungry, the instinct is to look for foods that will provide fast energy and raise your blood sugar level, such as high-carbohydrate snacks or sugary soft drinks. These quick-fix foods will not keep you feeling energetic or satisfied for very long—what they will do is add lots of empty calories.

Sleep Well to Lose Weight

A good night's sleep can also be an aid to metabolism. Researchers at Stanford University and the University of Chicago found in separate 2004 studies that sleep deprivation may be linked to obesity. In the Chicago study, participants were limited to four hours of sleep on consecutive nights. At the

end of the trial period, they were found to have lower levels of leptin, a hormone that suppresses appetite (▶ **48**), and higher levels of ghrelin, a hormone that triggers hunger. Participants had greater appetites, especially when it came to sugary foods. The Stanford study found that individuals who slept less than eight hours a night on average were heavier than those who got adequate sleep (adjusting for other variables).

Sleep deprivation has been suggested as a major contributing factor to today's obesity epidemic, according to a 2008 review published in the *European Journal of Endocrinology*. Adults typically require between seven and nine hours of sleep per night, so to keep your metabolism running smoothly, try to get at least seven hours per night. That will also help protect you against rises in blood pressure and blood sugar (▶ **84**).

27

Eat Oatmeal to Suppress Appetite and Boost Metabolism

Oatmeal can work better than diet pills to suppress appetite, according to Dr. David Katz, medical contributor to *Good Morning America*. That's because this natural source of soluble fiber breaks down gradually and stays in your intestines a long time, releasing nutrients slowly, which gives you a sense of fullness. Not only does a bowl of oatmeal for breakfast supply many essential nutrients, it can also provide day-long benefits to your metabolism.

Oatmeal is a whole grain. This means the oats from which oatmeal is made consist of the whole grain and its constituent components: endosperm, bran, and germ. Refined grains have gone through a milling process that removes the bran and germ. Whole grains are complex carbohydrates and digest more slowly than simple carbohydrates (such as glucose and fructose). One of the best sources of oatmeal is muesli, a Swedish cereal that's made of raw oats.

Get Your Metabolism Going with a High-Fiber Breakfast

Numerous studies, such as one published in a 2004 article in the *Journal of the American Medical Association*, have demonstrated the benefits of a metabolism-friendly diet containing foods like oatmeal that are heart-healthy and have a low glycemic-index (▶16).

Low glycemic-index foods are less likely to induce sharp spikes in blood sugar than foods high on the glycemic index. A half-cup of oatmeal contains five grams of fiber and only 75 calories (unless you load it up with sugar and milk). If you eat a breakfast filled with fiber, you'll be less likely to binge on doughnuts during your morning coffee break (▶28).

Eating a nutritious breakfast is one of the best ways to rev up your metabolism (▶26). Although a bowl of your favorite sweetened cereal may seem to offer quick energy, oatmeal provides far more health benefits, as a 2001 study published in *Nutrition Today* reports.

Protein and Whole Grains Provide Sustained Energy

Oatmeal also contains five grams of protein per serving, and protein gives you sustained energy that can be used for building muscle. Increasing lean muscle tissue is one of the best ways to raise metabolism, and experts advise eating protein with each meal for precisely this reason (▶17). Protein can also help regulate the amount of food you eat and how satisfied you feel.

Many of us can't stomach the thought of eating chicken or fish first thing in the morning, but the usual breakfast meats—such as ham, bacon, and sausage—are notoriously high in fat and sodium. Protein-enriched breakfast cereals and especially oatmeal offer a good alternative.

All oatmeal is not created equal, however. Whole-oat groats are said to provide the greatest health benefits because they've been minimally processed. Traditional oatmeal (steel-cut oats) is roughly chopped pieces of the entire groat (the oat kernel). Rolled oats are steamed and rolled before packaging. In instant oatmeal the oats have been further broken down for faster cooking; this process increases their glycemic index and therefore reduces their health benefits. Some

types of flavored instant oatmeal (particularly ones that are sweetened or labeled "creamy") can have twice the number of calories as plain oatmeal, with no added health benefit.

28

Lose Weight by Eating Fiber

High-fiber foods can facilitate weight loss by making you feel full longer. According to a 2008 study published in the *Journal of the American Dietetic Association,* thin subjects consistently ate more fiber than overweight subjects. When the diets of overweight and obese people were compared to those eaten by people of normal weight, results showed that people with higher body fat had consistently lower amounts of fiber in their diet. The Linus Pauling Institute, a micronutrient research center at Oregon State University, points out that fiber-dense whole grains aid digestion, which can contribute to weight loss.

Fiber is a self-regulating food for weight loss. As the name suggests, foods containing fiber tend to be fibrous—it requires more effort to chew them than it would to slurp down a milkshake. Having to slow down and chew food for a longer period of time gives your brain a chance to recognize that you are getting full. Studies, such as the above-cited one published in the *Journal of the American Dietetic Association,* have shown that people are less likely to overeat when they have to take time to carefully chew each bite.

Digesting High-Fiber Foods Requires More Energy

Metabolic rate increases after eating high-fiber foods, in part because of the additional work the body must do to digest them. Simple carbohydrates are comparatively easy to digest and, as such, do not challenge the metabolism to work harder. Complex carbohydrates, such as those found in whole grains, break down more slowly than simple carbohydrates, and high-fiber complex carbohydrates take even longer to digest (▶16).

Additionally, energy is released more slowly and continually during the digestion of high-fiber foods. Therefore, blood sugar levels remain more constant than when you intermittently eat low-fiber, high-sugar foods. A consistent blood sugar level can mean less snacking and more sustained energy.

Good Sources of Fiber

For anyone trying to lose weight and raise metabolism, fiber is an essential dietary component. While there is no recommended daily allowance (RDA) value for fiber, most experts suggest that adults consume twenty to thirty grams a day. Beans and lentils are some of the best sources of fiber. One cup of baked beans, pinto beans, white beans, or kidney beans contains sixteen to eighteen grams of fiber. Fiber-dense chickpeas (also known as garbanzo beans) contain metabolism-boosting calcium and magnesium as well. Broccoli is another big winner, with nine grams of fiber per cup. Grains such as buckwheat and bulgur—both have nine grams of fiber per cup, cooked—are also good choices. Corn contains ten grams of fiber per cup, chard and kale eight grams per cup each, peas provide eighteen grams per cup, and yams (with the skin) have six to seven grams.

Many fruits are also high in fiber. Figs have two grams of fiber each. Pears have four grams of fiber, apples contain three to four grams, blackberries provide around eight grams per cup, and a small orange has 1.5 grams. Bran, wheat, and of course fiber-enhanced cereals also contain a sizable amount of dietary fiber; as always, read labels carefully to be sure that you are consuming what you intended without any "hidden" ingredients. Be careful when introducing more fiber into your diet, as too much too quick may lead to bloating, cramps, and flatulence. Increase your fiber intake gradually over several weeks to allow your body time to adjust.

Soluble and Insoluble Fiber

Generally speaking, fiber comes from the parts of a plant that humans cannot digest. What your grandparents may have called "roughage," today we call dietary fiber, or the kind we eat (even though not all dietary fibers actually contain fiber). Dietary fiber is categorized into two basic types: soluble and insoluble. Insoluble fiber keeps your food moving through the digestive tract. Soluble fiber helps lower total cholesterol levels and can aid with regulating blood sugar.

Another way fiber helps you lose weight is by forcing you to drink lots of water (▶ **37, 38**). By its nature, fiber swells in the intestines. If you are taking fiber supplements or eating particularly fibrous foods, drink plenty of water to ensure that the fiber does its job.

29

Change Your Diet to Motivate Your Metabolism

Shake up your metabolism and get out of a weight-loss rut by changing your caloric intake. Sometimes eating a few *more* calories per day—temporarily anyway—may do the trick, as it stimulates your metabolism and shifts it into a higher gear. You can then switch back to a lower level after a few days to shake up your system yet again.

Try changing your diet so that for a few days more of the calories you eat come from protein (▶17)—this can motivate your metabolism. Studies, such as one cited in a 2008 article in the *Journal of Clinical Endocrinology and Metabolism*, show that you tend to feel less hungry when you incorporate more low-fat protein into your diet. Increasing your protein intake not only helps you feel full, it helps build muscle.

Consider other food substitutions as well. If you're sick of lettuce and carrots, replace them with broccoli and tomatoes for a couple days. New, nutrition-dense foods can revitalize your interest in weight loss while challenging your body to pick up the pace.

Challenge Your Body to Pick Up the Pace

Why do weight-loss plateaus occur? As you lose weight, your BMR, or basal metabolic rate (see p. 13), lowers correspondingly. Suppose a 150-pound (68 kg) thirty-five-year-old female has a BMR of 1,444. If she loses ten pounds (4.5 kg) before her next birthday, her new BMR will be 1,400. This takes into account the fact that this slimmer woman now needs fewer daily calories to maintain her weight than she did when she was ten pounds (4.5 kg)

heavier. Unless caloric intake changes, weight loss can periodically stagnate, and studies, such as a 1989 study published in the *Journal of Consulting and Clinical Psychology*, confirm the ease with which these plateaus can temporarily interrupt progress.

Our bodies can start to become lazy when they're not challenged. Although exercising daily and eating appropriately are sure-fire ways to lose weight and boost metabolism, the body has an amazing ability to adapt. If you lift the same weights in the same amounts every day, you won't continue to see great results. The same can hold true for caloric intake. Even though you might be eating the same foods that helped you lose weight before, you could find yourself in a weight-loss rut.

Work Your Way Out of a Rut

Another good way to get out of a weight-loss rut is to step up your exercise. Working out longer and harder can provide the emotional and physical stimulation required to raise energy and boost metabolism. Changing the type of exercise you do can also be effective.

A number of studies, such as one published in a 2007 article in the *Journal of Strength and Conditioning Research*, show that once you've adapted to a particular form of exercise, it's no longer as effective at building muscle as it once was. Other studies, such as one presented in an article from 2000 in the *Journal of Sport Behavior*, indicate that people tend to become less interested in exercise when they do the same thing every day. So, try going for a jog instead of riding an exercycle at the gym, for instance, or try sprint or interval training to rev up your metabolism and recharge your routine (▶8).

If you have lost a significant amount of weight quickly by going on a very low-calorie diet (which is not recommended), your metabolism may be suppressed to the point that losing more weight becomes difficult. If that's the case, you need to boost your metabolism to a higher level by slowly increasing your calorie intake back up closer to your BMR (see p. 13). You may see a weight increase at first, but over time, as your metabolism readjusts, you can begin to gradually reduce your food intake again to compensate. This will allow you to slowly lose weight while keeping your metabolism high.

During different times of our lives, our caloric needs can change. If you are forced to be more sedentary than usual—for example, while recovering from an injury—you will need fewer calories than when you were more active. If a job change has you on your feet for hours at a time, your caloric needs will increase. When in doubt, use the formulas on page 13 to figure out approximately how many calories you need per day to maintain your weight. From there you can figure out how many calories you'll need to cut to lose weight.

30 Eat Whole Fruits Instead of Juices

Eat your food, don't drink it. According to a January 2009 article published in the *Boston Globe*, most people get close to 20 percent of their daily calories from beverages. So think again before you reach for a bottle of apple juice or a raspberry smoothie. Your metabolism will get a bigger boost if you eat whole fruits instead, because they take significantly more calories to digest than liquid or processed fruit (such as applesauce).

Juices are already broken down, therefore they require very little energy to digest; moreover, the sugar in juice provides only a brief metabolic boost. In contrast, "nutrient-dense" whole fruits and vegetables must be broken down by the body's digestive processes over a period of time, thereby expending more energy and burning more calories.

An Apple a Day to Keep Weight Away

Additionally, many fruits—including apples—contain chemicals called polyphenols. These antioxidants include antiobesity properties that can aid weight loss, according to the *Journal of Personalized and Systems Medicine*. These fruit extracts can also lower blood cholesterol and glucose levels. Whole fruit ranks lower than juice on the glycemic index, too. A 2004 study published in the *Journal of the American Medical Association* showed that reducing your glycemic load, by consistently eating foods that are low on the glycemic index, can help regulate metabolism and promote weight loss.

Aside from the calories burned during digestion, whole fruits often contain fewer calories than their juice counterparts. A cup of apple juice contains about 120 calories and has no fiber; by contrast, a medium-sized apple has only eighty calories and provides nearly four grams of fiber, which comes from the skin and pulp. Obviously, you'll spend more time and burn more energy eating the apple than drinking apple juice. Plus, the fiber in the apple will cause you to feel more full (▶28). An orange contains an average of sixty to seventy calories and four grams of fiber. If you drink an eight-ounce (240 ml) glass of orange juice, you'll ingest about 120 calories—double the calories you'd get from eating the whole orange.

Choose Low-Calorie, High-Fiber Fruits

So far we've been comparing apples and oranges. The U.S. Department of Agriculture's Nutrient Database shows

that while eating fruit is one of the best ways to add fiber to your diet, not all fruits are created equal. An eight-ounce (240 ml) glass of grape juice, for instance, contains about 155 calories, whereas the same amount of grapefruit juice only has about ninety calories. Dried fruits provide plenty of fiber, but they also have more calories than fresh, whole fruit. A quarter-cup of raisins contains sixty calories more than an apple. A 3.5-ounce (100 g) serving of dried apricots gives you eight grams of fiber, but almost 240 calories—five times more than a fresh apricot. For metabolism-boosters, fresh raspberries offer the best of both worlds—one cup provides more than eight grams of fiber, but only sixty calories.

There are, of course, times when pro-cessed or liquefied versions of whole foods are beneficial and convenient. Juices can be consumed quickly, pro-vide fast energy, and may be better tolerated during illnesses—when medical conditions prevent the chewing or swallowing of solid foods, liquid nutrients become a necessity. And undeniably, fruit or vegetable juice is a healthier choice than soda. But in most situations whole fruits will provide more metabolic bang for the buck than juice.

31

Soup It Up to Take Weight Off

Some like it hot—soup, that is. John Foreyt, Ph.D., Director of the Behavioral Medicine Research Center at Baylor College of Medicine, Texas, says, "People tend to eat less following a bowl of hot, nutritious soup. It's usually hot, so you eat it slowly and feel fuller."

There are distinct metabolic advantages to starting a meal with a liquid (▶38). However, that liquid needn't be water. Beginning a meal with a broth-type soup can have the same appetite-suppressing effect as drinking a large glass of water before eating. To get the most beneficial effect, wait fifteen to twenty minutes before moving on to the next course.

Studies, such as one published in a 2007 article in the journal *Appetite*, show that starting a meal with soup can result in consuming up to 20 percent fewer calories during the course of the entire meal. In a study conducted at Pennsylvania State University, people who started lunch with soup reduced their subsequent caloric intake by an average of 100 calories.

Dr. Melina Jampolis, physical nutrition specialist and CNNhealth.com's diet and fitness expert says, "Don't forget to eat lots of water-based foods like soups … which are equally important for weight loss, as they lower the calorie density of meals. That can help you reduce calories without reducing portions."

Another study published in a 2005 article in the journal *Physiology & Behavior* suggests that soups generate high feelings of satiety; therefore, making a meal out of a soup can fill you up without packing on calories. Low-sodium vegetable soups (which usually contain about seventy calories per cup) and fat-free chicken broth (forty calories a cup) are good choices if you're trying to lose weight. Black bean and pasta-free minestrone soup have about 125 calories per cup, but they're loaded with fiber—and digesting fiber burns up lots of calories (▶28).

What Soups Are Best for Heating Up Your Metabolism?

If your goal is raising metabolism, you should aim to eat healthy fats, including omega-3 fatty acids (▶21), rather than consuming soups like clam chowder (which contains about 300 calories per cup) that are chocked full of cream and other saturated fats.

Back in the 1980s, cabbage soup gained popularity as a dietary aid. This low-calorie, fat-burning "filler food" has been shown to reduce hunger pangs and cravings that can lead to consuming excess calories during a meal. However, a diet consisting mainly of cabbage soup should only be undertaken for a short period of time, say a day or two at most.

Spicy soups provide the metabolism-boosting benefits of what's known as "thermogenesis" (▶19). According to a 2006 article in *Physiology & Behavior*, spicy foods give off heat that helps the body burn fat for up to a few hours after you've finished eating. To get the same thermic benefits, you can spike up plainer soups with Worcestershire sauce, cayenne pepper, ginger, or hot mustard.

Chicken soup is a classic "comfort food," something you eat when you're sick, right? However, diets based on chicken soup also show impressive weight-loss results—so long as the diet only lasts a short time. Low in fat and rich in nutri-ents, protein, and liquid, a small portion of chicken soup can help you feel sated and nourished, so you're less inclined to eat a high-calorie meal afterward.

Eat a Meta-Boost Breakfast to Burn Calories All Day Long

Did your mother tell you breakfast was the most important meal of the day? Guess what, she was right. Don't start the day on an empty stomach. Studies, such as a 2005 article published in the *Journal of the American Dietetic Association*, confirm that people who eat a healthy breakfast tend to lose more weight than people who skip breakfast.

"When people skip breakfast," says U.S. Department of Agriculture researcher Shanthy Bowman, Ph.D., "they end up eating more calories by the end of the day, and we know that they end up compensating for this skipped meal with high-sugar, high-fat foods."

Your morning meal helps rev up your metabolism after its relatively slow period overnight, and this first intake

of calories is what sets the tone for your entire day (▶26). If you are tempted to skip breakfast and wait until lunch to eat, you're making a big metabolic mistake by missing out on an opportunity to burn more calories at a higher rate all morning. Additionally, a study published in 2007 in the medical journal *Metabolism* found that skipping meals early in the day could lead to overeating at dinner, which produced detrimental changes in the metabolisms of test subjects.

Jump-start Your Metabolism in the Morning

However, "breakfast" doesn't mean binging on doughnuts or biscuits and gravy every day. The key to raising your metabolism—and keeping it high throughout the day—is making good

food choices at breakfast. The first thing you consume in the morning should be a big glass of water, which will help hydrate you after an night's sleep and can help you feel full enough to prevent overeating (▶38). Drink a cup of water before you start preparing breakfast, then aim to start eating about ten to fifteen minutes later.

Cereal is a traditional breakfast food, and the right type can boost metabolism and enhance weight loss. One of the best ways to start your morning is with a bowl of steel-cut oatmeal (▶27). These powerhouse oats contain 160 calories per serving, three grams of fat, eight grams of fiber, and six grams of protein. They can sustain you through the morning while feeding your muscles the protein they need to grow stronger.

If you're really crunched for time, a bowl of cold, prepared cereal is fine—but choose one that's high in fiber, made of whole grains (▶16, 28), and low in both sodium and sugar. And make sure you use non-fat milk to gain the extra benefits of calcium and amino acids (▶18, 23). Whole-grain toast, English muffins, or bagels provide complex carbohydrates that will break down slowly and continue to give you energy throughout the morning.

Dairy Products Provide Muscle-Building Protein

Non-fat yogurt is another good breakfast choice. Most plain yogurt contains approximately 100 calories per eight-ounce (225 g) serving, and provides about five grams of muscle-building protein. Yogurt is also an excellent source of calcium (▶23). Look for varieties that contain active cultures and/or probiotics, which can aid digestion and enable you to get the most out of what you eat.

Rather than reaching for flavored, presweetened yogurt, try topping some plain yogurt with a handful of raspberries, which provide lots of fiber but few calories (▶30). If you like, add a little wheat germ to gain additional whole-grain benefits as well as metabolism-boosting protein.

Despite what you've heard about eggs and cholesterol, eggs are traditionally part of a hearty, low-calorie breakfast and a great source of protein. One hard-boiled, poached, or scrambled egg provides about six grams of protein and only ninety calories.

What about that wake-up cup of coffee or tea? Studies, such as a 1998 article in the *Journal of Applied Physiology*, confirm what most of us know experientially: the caffeine in coffee and tea can boost your metabolism in the morning (▶40). If you prefer, substitute green tea to not only raise your metabolism but also burn fat (▶39).

33

Feast on a Meta-Boost Lunch to Keep Your Metabolism High

Get off the blood-sugar roller coaster by eating a high-fiber, high-protein lunch. Eating a heavy, high-calorie meal at noon can lead to a "food coma" slump a few hours later. If you've eaten a hearty, healthy breakfast (▶32), you may not need to eat a lot at midday in order to keep your metabolism running smoothly.

Before you begin eating lunch, drink a large glass of water—this helps you rehydrate and prevents thirst signals that may be masquerading as hunger (▶38). Remember to pace yourself as you eat—eating slowly (▶68) allows your body to relay "full" signals to the brain, so you're less likely to overeat. Slowly eat a portion of your meal (say half) and then stop to drink another glass of water. If you're really still hungry roughly fifteen minutes later, eat half of what's left on your plate and then reevaluate—you might find you don't need to clean your plate completely.

Go for the Greens
Salad can be an excellent choice for a midday meal, because lettuce and other salad greens are very low in calories. Let the shade of green guide you: Dark green leaves such as those of romaine and spinach are better sources of antioxidants, folic acid, and other nutrients than pale iceberg lettuce.

If a salad bar offers lots of add-ons, skip the bacon bits and croutons; instead, choose vegetables or fruits that are high in immunity-boosting antioxidants (such as tomatoes) and/or fiber (▶24, 28) Asparagus, broccoli, lettuce, spinach, zucchini, and green beans, for instance, are low in calories and also require lots of energy to digest, so you'll come out ahead in the calorie-counting game. If you prefer, try a fruit salad with a cup of strawberries (fifty calories, three grams of fiber) plus a half-cup of non-fat cottage cheese (100 calories and fourteen grams of protein).

Don't make the mistake of loading up your salad with gobs of dressing. Two tablespoons of creamy ranch or Caesar dressing can add 150 to 200 calories—many from saturated fat—to your weight-wise salad. According to a 2004 study published in the journal *Diabetes Care*, vinaigrette dressing made with red or white wine vinegar may have the additional metabolism-regulating benefit of lowering blood sugar for some people (particularly diabetics).

Replenish Your Metabolism with Protein at Lunch

You should avoid fat-laden burgers for lunch, but a sandwich made with lean meat can provide a metabolism boost at midday. Protein requires more energy—and more calories—to digest than carbohydrates or fats (▶17). The extra energy your body must expend to digest protein results in a temporarily increased metabolic rate. Protein also keeps you feeling full longer, so you're less likely to snack between meals.

A three-ounce (85 g) serving of lean processed turkey contains around eighty calories, one gram of fat, and fourteen grams of protein. A lean serving of beef sirloin is roughly double in both calories (160 calories) and protein (twenty-five grams), though significantly higher in fat (five grams). Two ounces (57 g) of cheese add about 230 calories and twenty grams of fat. When it comes to the bread, dark, whole-grain breads are the best choice. High in fiber, these complex carbohydrates take longer to digest, meaning that the sugars are absorbed slowly and your body is less likely to store any excess as fat.

34

Keep Metabolic Flames Burning with Meta-Boost Snacks

Spread out your daily intake of calories to keep your metabolism running at a high rate. Experts suggest you're better off eating five small meals, or three meals and two substantial snacks, rather than the usual three large meals per day.

As Dr. Melissa Dawahare, a licensed primary care physician in Tempe, Arizona, explains, "Eating five smaller meals throughout the day minimizes the high spikes and low drops in glucose and insulin levels. In addition to avoiding symptoms of hypoglycemia, smaller meals help to decrease cravings or mood swings. Smaller meals also help you to avoid overeating at mealtime." If you consume more calories during a meal than your body needs for energy, the result can be increased fat storage and a sluggish metabolism.

Eat Snacks to Lose Weight

How many calories should a mid-afternoon snack contain if you're trying to lose weight and raise metabolism? The numbers will vary for each individual based on the individual's height, weight, and exercise routine, but for the average person a 1,500-calorie-a-day diet should be sufficient for losing weight at a rate of about a pound (0.45 kg) per week.

Dividing this by five (three meals and two snacks) and you get about 300 calories per meal or snack. Most of the time, breakfast will end up on the lower end of the scale and dinner can be a little higher, so you may be closer to 200 calories for each snack. Snacks that are high in protein and fiber can actually help reduce calories, and turn snack time into a metabolism-benefiting event.

Before eating any snack, always drink a large glass of water. Most of us don't drink enough water, and it's easy to confuse hunger with thirst signals (▶37, 38). If you're still feeling hungry fifteen minutes afterward (and it's been two hours or more since your last meal), you're probably legitimately hungry.

Snacks that Combine Fiber and Protein

For a real metabolism-kicker, try a bowl of spicy salsa topped with a small serving of shredded, low-fat cheese. The thermogenic properties of hot fresh salsa, as shown in a 2005 study published in the journal *Public Health Nutrition*, raises metabolism (▶19); the cheese will add muscle-building protein and healthy amino acids. Instead of corn chips, dip sliced green peppers or celery into the salsa mixture for added crunch.

Consider a snack that combines low-calorie vegetables with muscle-building protein. A celery stalk with a tablespoon of peanut butter contains about 100 calories, eight grams of fat, one gram of fiber, and four grams of protein. Cut down on fat by switching to low-fat peanut butter, and increase both the fiber and protein content by sprinkling wheat germ on top.

If you're hankering for something warm and filling, try a pack of instant unflavored oatmeal—it has just 100 calories but contains 2.5 grams of fiber and 3.5 grams of protein. A sprinkle of brown sugar or a few raisins can turn it into a sweet snack without adding lots of fat. Cooked rhubarb, strawberries, or blueberries also make terrific mix-ins.

Soup Can Be an Ideal Metabolism-Boosting Snack

Soup helps generate high feelings of satiety (▶31)—it fills you up without packing on calories, as indicated by a study published in a 2005 article in the journal *Physiology & Behavior*. Chicken broth has around forty calories per cup—toss in a handful of sliced carrots, scallions, or peas to add interest without adding many calories. Tomato soup has around seventy calories per cup, and a light chicken vegetable soup has around eighty calories per cup. Soups can fill you up without weighing you down; just avoid cream-of-anything varieties, or ones filled with white pasta, which can add unwanted calories and fat.

If you need an on-the-go metabolism-raising snack, try an old favorite: trail mix. Make a mixture consisting of equal parts walnuts (high in omega-3 fatty acids, which can aid weight loss, ▶21), almonds (to help control blood sugar), and dried cranberries (low in calories and full of antioxidants). Add relatively small amounts of other dried fruits, sunflower seeds, and a few dark chocolate chips. A small handful will provide around 200 calories and give you sustained energy.

With a little bit of foresight and planning, snacking doesn't need to destroy your weight-loss and metabolism goals. On the contrary, healthy snacks eaten in small portions throughout the day can help keep your appetite under control while revving your metabolism and providing a steady supply of energy.

35

Eat Light When You Eat Late

For many people, dinner is something of a social event, but it can also be an opportunity to pack on weight. Instead of eating your biggest meal at suppertime, eating light at the end of the day makes better metabolic sense. According to Shanthy Bowman, Ph.D., a researcher for the U.S. Department of Agriculture, "In America, we eat more during dinner than any other meal." And a USDA survey showed that overweight adults tended to consume significantly more calories at suppertime than adults of normal weight.

Eating fewer calories as you get closer to bedtime makes sense physiologically, as sleeping requires less energy than your daytime activities. However, in a study of eating patterns, John de Castro, Ph.D., chairman of the psychology department at the University of Texas at El Paso, found that on average people consumed 42 percent of their total daily calories during and after dinner.

Plan to eat your evening meal at least two to three hours before bedtime, and keep it light—you'll sleep better. An article in the *Bulletin of European Shift Work Topics* suggests that eating a heavy meal can make it harder to go to sleep—and lack of sleep has been linked with obesity, as researchers at Stanford University and the University of Chicago documented in 2004 (▶ 26).

Low-Cal, Metabolism-Raising Meals

An ideal meal for raising metabolism combines protein (for building muscle), healthy fats (for regulating blood sugar metabolism), and complex carbohydrates (for fueling the body with energy-laden nutrients). Try to create meals that combine all these elements, and pay careful attention to both calories and portion size.

A three-ounce (85 g) serving of salmon, for example, has about 175 calories, nineteen grams of protein, and lots of omega-3 fatty acid (▶ 21). An equivalent portion of lean sirloin steak contains about the same number of calories and slightly less protein, but no omega-3s. Therefore, a meal that includes a serving of baked or broiled salmon, a cup of steamed broccoli (thirty calories, two grams each of fiber and protein), and a half-cup of brown rice (100 calories, plus plenty of complex carbs) provides good nutrition while meeting the hallmark attributes of a metabolism-raising meal.

Vary your protein source to keep your routine interesting: lean pork, chicken, turkey, beef, and fish are all good choices. Skinless roasted chicken weighs in at about 100 calories per three-ounce (85 g) serving, whereas a standard hamburger will pack on 430 calories (with the bun).

Vegetarians should include protein sources such as cooked lentils (230 calories, eighteen grams of protein, fifteen grams of fiber per cup) or soybeans (375 calories, thirty-three grams of protein, ten grams of fiber per cup). Tofu (seventy calories, six grams of protein per three-ounce [85 g] serving) makes an excellent main ingredient for a vegan stir-fry; combine it with low-calorie, high-nutrient vegetables such as zucchini, carrots, cabbage, or asparagus.

Consider adding complex carbs in the form of baked sweet potatoes, brown rice, or whole-grain bread. Sweet potatoes fall lower on the glycemic index than white potatoes, meaning they're less likely to leave you with a spike in blood sugar levels—and they're low in calories (about 100 per medium potato), high in fiber (four grams), and a great source of potassium (550 mg).

Instead of butter or sour cream, top your potato off with spicy salsa to increase metabolism—jalapeño, habanero, and other hot peppers can raise your metabolism by increasing your body temperature (▶19). However, spicy foods can cause heartburn in some people, so if you are susceptible to heartburn save this metabolism-boosting tip for earlier in the day.

Drink Water Rather Than Wine at Dinner Time

Although a glass of wine with dinner may be tempting, wine is high in calories (around eighty to 100 calories per four-ounce [120 ml] glass, twice that for port); some other alcoholic drinks contain even more calories (▶80). In addition, according to a 1997 study published in the journal *Nutrition Reviews,* alcohol can slow the rate at which you metabolize fat.

Drinking a full glass of water before starting your meal can reduce hunger symptoms and cause you to eat less (▶38). But large amounts of fluids taken close to bedtime can also mean more trips to the bathroom during the night; try to "front-load" most of your fluids toward the first part of the meal.

36

Choose Desserts that Won't Wreck Your Metabolism

Many of us like to have a bite of something sweet after the final meal of the day. But the truth is, when trying to lose weight and raise metabolism, the best dessert is usually no dessert at all. There is no physiological reason to consume extra calories right before you go to bed, considering that your body requires fewer calories while sleeping. A relaxing cup of caffeine-free tea may help prepare your body and mind for bed better than an after-dinner cordial and a piece of chocolate layer cake.

Elaine Magee, M.P.H., R.D., the "recipe doctor" with the WebMD Weight Loss Clinic suggests, "If you're in the habit of finishing your day with dessert, try having a mini-portion. The first few bites of a food always taste the best, anyway. Experts say a petite portion is more likely to satisfy if you choose a dessert you truly enjoy, take your time, and savor every bite, and accompany your treat with a cup of hot coffee or tea." A single piece of dark chocolate, for instance, only contains about thirty calories and may be enough to satisfy your sweet tooth.

Fruit Provides a Nutritious, Negative-Calorie Dessert

Consider eating fresh or frozen fruit as an after-dinner treat. Some fruits, including apples, berries, melons, and pineapple, fall into the "negative-calorie" category, meaning that they burn up more calories during digestion than they provide (▶24). A plate of mixed berries, served with a side of non-fat yogurt, makes a visually appealing, weight-friendly meal finale.

You might want to include pineapple in your fruit bowl, as pineapples have enzymes such as bromelain that can assist with digestion. Alternatively, if you're craving something cold and creamy, many all-natural fruit sorbets have only about 100 calories in a half-cup serving and are fat-free, yet can be delicious and satisfying.

Vegetable and fruit breads—zucchini, pumpkin, cranberry, banana—are appetizing and healthy choices, too. A slice of zucchini bread contains approximately 190 calories, and if made with whole-grain flour provides some complex carbohydrate benefits as well. Cranberry bread weighs in at around 170 calories—about half what you'd get in a slice of cheesecake—and gives you antioxidants as well.

You've probably heard the saying "from soup to nuts," right? Many of our grandparents served nuts as an after-dinner treat—and research has shown that they made a healthy choice. Nuts contain protein and omega-3 fatty acids. You can even dress up walnuts and almonds with a drizzle of honey, or sprinkle them with brown sugar and cinnamon before roasting.

Bear in mind that when it comes to desserts, almost anything is fine when eaten in moderation. One bite of Boston cream pie will not make you fat. However, many of us lack the willpower to stop at a single bite, and a slice of chocolate cake can set you back at least 250 calories. Use common sense to make intelligent dessert choices, ones that you won't regret in the morning.

PART III

Drink to Your Success:
Metabolism-Friendly Beverages

31

Drink Water to Increase Your Metabolic Rate

Drinking a few extra glasses of water a day can boost your metabolism. Research conducted in 2003 at the University of Utah demonstrated that dehydration leads to a slower metabolism. Another 2003 study, reported in the *Journal of Clinical Endocrinology & Metabolism*, found that thirty minutes after drinking about two cups of water, test subjects averaged a 30 percent increase in their metabolic rates. Interestingly, for the men in this study the increase in metabolism resulted from an enhancement of fat burning, while for women it came from an increase in carbohydrate breakdown.

As we all know, water is essential for life. The human body is composed of about 60 percent water. Drinking water before eating, according to studies summarized in a 2007 article in the magazine *Consumer Reports*, suppresses your appetite and fills you up, so you don't need to eat as much to feel satisfied (▶38). Water also assists in the smooth movement of food through your system; it aids the breakdown of food particles, encourages the absorption of nutrients, and helps the body eliminate waste.

How Much Is Enough?

How can you tell if you're drinking enough water? One of the first signs of dehydration is thirst. However, by the time you feel thirsty or notice a dry, sticky mouth, you're already slightly dehydrated, so it's a good idea to drink water throughout the day. A general guideline is to take in at least eight eight-ounce (240 ml) glasses per day, plus an additional eight ounces (240 ml) for every half-hour you exercise. Jens Jordan, M.D., recommends drinking a total of eight to twelve cups daily. Cool water appears to provide the best results if your goal is to lose weight because the body burns calories to warm the liquid to body temperature (▶38).

If that sounds like a lot, consider that about 25 percent of your total daily water consumption can come from the food you eat: cucumbers, eggplant, tomatoes, stone fruits, and citrus fruits all have a very high water content. Drink enough to keep your urine the color of light lemonade—dark urine can be a sign of a water deficit.

Other signs of dehydration include joint pain, headaches, constipation, and dizziness. You can avoid getting to that point by drinking water regularly throughout the day to speed your metabolism and encourage weight loss, in addition to other health benefits.

What About Other Beverages?

You can also count other beverages you drink as part of your daily intake. Juice, herbal tea, and soft drinks all provide fluid. When drinking liquids specifically to raise metabolism and lose weight, though, remember that not all drinks are created equal. Soda and juice contain sugar and calories, which may not be desirable. Also be careful with caffeinated drinks that act as diuretics and increase urination. Coffee and caffeinated tea can actually cause the body to dehydrate by flushing fluids from the body—drink a glass of water with your cup of coffee in the morning. Alcohol can also be dehydrating, and it's high in calories—these two factors tend to slow metabolism rather than raise it (▶80). Plain and simple water should be your primary beverage of choice.

Bear in mind, however, that it is possible to drink too much water. Your water intake needs to be balanced with salt and other electrolytes. Hyponatremia, a condition that occurs when the body's salt stores are overly diluted by too much water, is infrequent but potentially fatal. If you are not used to drinking much water, build up gradually to your optimal intake level.

Fill Up without Filling Out

Drink up to fill up, without ingesting any calories. Studies, summarized in a 2007 article in the magazine *Consumer Reports*, confirm that drinking water before eating suppresses your appetite and can help you eat less. A 2007 paper published in the journal *Obesity* confirms that consuming a beverage before a meal can significantly reduce the amount of food consumed at mealtime—particularly for older adults who already have slower metabolic rates than their younger counterparts.

Raise Metabolism and Lose Weight by Drinking Water

A study conducted by Dr. Brenda Davy, Associate Professor of Human Nutrition, Foods, and Exercise at Virginia Tech, published in a 2008 article in the *Journal of the American Dietetic Association*, showed that people who drank water before eating a meal saved themselves about seventy-five calories. Over a year's time, that can add up to a net loss of more than fourteen pounds (6.4 kg). According to CNNhealth.com's diet and fitness expert, Dr. Melina Jampolis, Dr. Davy also found that people who drank two glasses of water twenty to thirty minutes prior to every meal lost more weight and lost it faster than their counterparts who didn't drink water.

A study performed at Berlin's Franz-Volhard Clinical Research Center and reported in a 2003 issue of the *Journal of Clinical Endocrinology & Metabolism* showed that shortly after test subjects drank a half liter of water their metabolic rates increased by 30 percent. Researchers calculated that simply by drinking an additional 3.2 pints (1.5 l) of water per day for a year, you'd burn 17,400 calories and lose about five pounds (2.3 kg). Interestingly, up to 40 percent of the extra calorie-burning cited in this study was estimated to be due to the body's effort to heat up the cold water. Drinking cold water forces your body to work harder to keep your body temperature normal; the cooling process burns calories. Whenever your body is subjected to significant temperature changes—hot or cold—it must exert extra effort to maintain your metabolic rate (▶13). Therefore, drinking cold water before meals offers a two-for-one boost to your metabolism.

Are You Really Hungry, or Just Thirsty?

Often, our brains play tricks on us by sending mixed messages about hunger

and thirst. It's easy to confuse hunger and thirst. If you think you're hungry, but you've eaten within the last couple of hours, there is a good chance that you're really thirsty instead. In fact, by the time you actually feel thirsty, your body is already slightly dehydrated. Rather than reaching for a snack, drink a glass of water, then wait ten minutes and see if the urge to eat is quelled.

When you eat, your body releases proteins that send messages to your brain telling you to stop eating. Because it takes about twenty minutes for you to realize that you're full, as shown by a 1980 study published in the journal *Science*, it's easy to overeat if you're not paying attention to those "full" signals.

Drinking a glass of water before eating accelerates the feeling of fullness—without adding a single calorie. That's because water reaches your stomach quickly. Additionally, a glass of water takes up room in your stomach, which means you don't have to eat as much to feel satisfied. Once you've silenced those hunger pangs, you're in a better position to make sensible choices about the size of your meal. Next time you go out to eat at a restaurant, drink a glass of water before you look at the menu—you'll avoid ordering more than you really need to eat.

39

Drink Green Tea to Burn Fat

Can simply drinking a soothing cup of green tea in the morning metabolize fat? Research done in the East as well as in the West suggests the answer is *yes.* Cell culture studies have found that green tea prevents fat cells from growing and multiplying. Green tea also inhibits the enzymes involved in the fat-storage process, and supports the enzymes involved in the fat-oxidation process. That's a winning combination for people who want to burn fat without exerting much effort. However, if you drink green tea *and* exercise, you may experience increased fat breakdown, according to an article published in 2008 in the *American Journal of Clinical Nutrition.*

A 2005 study published in the *American Journal of Clinical Nutrition* showed that men who drank tea enriched with green tea extract had a significant decrease in their obesity markers including BMI and abdominal fat. Dr. Nagao Tomonori and his colleagues from the Kao Corporation in Japan conducted a small trial to find out whether green tea really can help burn fat. Over a period of twelve weeks, test subjects drank either green tea or oolong tea at supper. The green tea was enriched with 690 milligrams of catechins (antioxidants found in green tea, as well as in some cocoas). The oolong tea served as the "control" contained only 22 milligrams of catechins. At the end of the study, the men who had consumed green tea had a total fat level 7.5 percent lower than that of their colleagues in the control group. Their body weight was 1.5 percent lower and their average waist measurement was 2 percent smaller.

What's Behind Green Tea's Metabolism-Boosting Benefits?

Other studies, such as one published in a 1999 article in the *American Journal of Clinical Nutrition*, suggest that green tea may cause metabolic rate to rise, partly because it makes the body work harder and expend more energy. This increase could be due to the caffeine in green tea. Green tea can have twenty to fifty milligrams of caffeine per cup; coffee, by comparison, has between eighty and 180 milligrams per cup.

For someone who's trying to raise metabolism, caffeine can offer advantages (▶40); but if you're trying to watch your caffeine intake, consider drinking decaffeinated green tea—research suggests you'll still get the benefits, including an increase in metabolism.

A study carried out in 1999 by Dr. A.G. Dulloo of the University of Geneva indicated, however, that an antioxidant known as epigallocatechin gallate (EGCG)—not caffeine—may be what's responsible for the notable metabolism boost. Green tea contains more of these antioxidants than other types of tea because its leaves, unlike those of black teas, are not fermented. Dr. Dulloo's study compared the effects of green tea to those of other caffeinated teas. The results showed that subjects who consumed ninety milligrams of EGCG experienced a boost in energy; control subjects who ingested the same amount of caffeine (derived from other sources) enjoyed no such metabolic effects.

The University of Chicago's Tang Center for Herbal Medical Research also studied the effects of the epigallocatechins in green tea, and discovered that drinking green tea offered additional health benefits for people who want to lose weight. Specifically, it helped lower blood levels of glucose (sugars), lipids (fats), and cholesterol.

The Linus Pauling Institute says this effect of green tea on weight loss is "apparently due to increased fat oxidation and thermogenesis." Thermogenesis refers to energy used in heat production that is not related to resting metabolism or physical activity. Diet-induced thermogenesis is related to energy-consuming processes involved in the digestion of a particular food or drink (▶19). Simply put, the body, when exposed to heat or cold in the form of food or environmental circumstances, attempts to regulate its temperature— and burns calories in the process.

How Much Is Enough?

How much green tea do you need to drink to reap its benefits? There is no recommended daily allowance (RDA) for green tea, and a wide range of opinions exists as to how much green tea should be consumed daily. Some sources say ten cups a day provide more health benefits than three cups a day, but others think that three to four cups a day is a smart and safe amount.

A 1999 Hong Kong study found that three cups of green tea a day significantly reduced blood levels of cholesterol and triglycerides. Triglycerides are the most common types of fat that exist in your body. Triglycerides in plasma are derived from fats in the food you eat; they can also be made in the body from other energy sources such as carbohydrates. Most calories you consume that are not used at once by your tissues are converted to triglycerides and transported to fat cells to be stored.

You can obtain the metabolic benefits of green tea even without brewing and drinking the tea itself. Green tea extract is available in both capsule and liquid forms. It's also added to a variety of foods and drinks, including yogurt, ice cream, energy drinks, and even some breakfast cereals.

Several medical conditions may preclude drinking green tea. For example, green tea may lessen the effects of chemotherapy drugs, and it can also change the way drugs such as lithium, adenosine, and beta-blockers work in your system. Green tea can also counteract certain medications, such as ephedrine and many cough/cold medications. Always check with your physician or specialist before trying any supplement, especially if you're already taking medication.

103

40

Jolt Your Metabolism with Caffeine

Are you someone who can't get going in the morning without a cup of java? Studies show that coffee not only helps you wake up and clears your head when you're feeling sluggish, it jump-starts your metabolism and burns fat. The force behind coffee is, of course, caffeine. As shown by a study of healthy young men, reported in the *American Journal of Clinical Nutrition* in 2004, "Caffeine ingestion increased energy expenditure 13 percent and doubled the turnover of lipids." Higher fat-burning potential, combined with a muscle-building exercise regime, is one of the best recipes for improving the rate of your resting metabolism.

Even if you're not a coffee drinker, you can get fat-burning benefits from other drinks that contain caffeine (▶ **39, 41**).

A study carried out in 2001 and published in the *Journal of Nutrition* found that test subjects who consumed full-strength oolong tea experienced a 12 percent increase in fat oxidation after drinking the tea, as compared to people who drank water.

Can Caffeine Improve Your Exercise Performance?

Some weight lifters believe that caffeine can combat muscle fatigue and alleviate pain associated with strenuous workouts. A study by University of Illinois community health professor and competitive cyclist Robert Motl, published in 2009 in the *International Journal of Sport Nutrition and Exercise Metabolism,* shows that caffeine reduces pain while exercising. Furthermore, a 1998 study published in the *Journal of*

Applied Physiology determined that ingestion of caffeine could increase exercise endurance.

Specifically, caffeine stimulates the central nervous system. Stimulants are a class of drug that acts to temporarily increase heart rate, body temperature, and your general sense of alertness. Because caffeine crosses the blood-brain barrier, it can make you feel mentally alert as well as physically energized. This effect typically lasts until the caffeine is completely metabolized, a process that can take a few hours. For this reason, some athletes consume caffeine prior to a workout.

A xanthine alkaloid like that found in yerba mate (▶ **41**), caffeine is found naturally in a variety of plants, beans, and nuts. Although coffee and tea are the most familiar sources of caffeine, you'll find it included in many energy drinks, sodas, chocolate, and flavored ice creams. Caffeine tolerance levels vary for individuals. Some people feel like they're getting enough caffeine from a cup or two of coffee a day, while others may require large amounts of cola or coffee to get the same energy boost or to ward off the side effects of caffeine withdrawal. If you are using caffeine to help you get more out of your workouts, it is best to consume caffeine right before a workout and to abstain from using it on a regular basis in order to keep your sensitivity to caffeine high.

Caffeine Side Effects

Because caffeine acts as a stimulant on the central nervous system, it is classified as a drug. And, like many drugs, caffeine is addictive. As your body gets used to having steady infusions of caffeine throughout the day, the metabolic benefits can decrease and you may need to drink more tea or coffee to get the same energizing effect.

When consumed in excess, caffeine can cause adverse side effects such as headaches, shaky or jittery limbs, restlessness, blood-pressure changes, irregular heartbeats, and sleep disturbances. Caffeine intake of more than 500 to 600 milligrams a day has been known to cause headaches. If you're pregnant or suffer from a disease for which caffeine is contraindicated (such as gastro-esophageal reflux, hypertension, or ulcers), talk to your physician or specialist before consuming any amount of caffeine.

41 Give Your Metabolism a Mate Boost

Can't drink coffee or tea? How about starting your morning energy-raising routine with something different? South Americans enjoy the health benefits and metabolic boost they get from a drink called mate. A 2005 study published in the journal *Economic Botany* showed that some people who were sensitive to caffeine tolerated mate-based drinks better than chocolate or coffee.

Avoid Caffeine's Side Effects

A brew made with the leaves of the yerba mate plant, mate derives its metabolism-boosting properties from the xanthines present in the tea. Xanthines are alkaloids, chemical compounds found in nature. Mate tea does have a small amount of caffeine, but less than coffee or black tea, and instead contains a similar substance called mateine. In

most people this works as a stimulant and supplies extra energy, without producing some of the negative side effects commonly associated with coffee or black tea, such as shaking hands and restlessness. Additionally, mate has been shown to act as an appetite suppressant.

How do mate's stimulating effects compare to what you'd get from coffee or tea? It has been estimated that a cup of mate tea has the rough equivalent of about fifty milligrams of caffeine, which puts it on par with green tea (▶39).

Other Metabolism-Boosting Benefits of Mate

Like green tea, mate tea may offer other benefits for metabolism-boosters as well. In 2005, a study of twenty-five types of mate performed at the University

of Illinois found it contained "higher levels of antioxidants than green tea." Additionally, yerba mate offers up to "90 percent more metabolism-boosting nutrients than green tea."

A 1999 study published in the journal *Phytomedicine* suggested that mate has thermogenic properties, meaning that it encourages the body to burn more calories as food is being digested (▶19). Mate is said to increase fat oxidation for the same reason. A 2007 study in the journal *Planta Medica* found that mate may help lower "bad" cholesterol (LDL) and increase "good" cholesterol (HDL). More study is, however, needed to ensure mate's safety, confirm its efficacy, and find the best "dose" with the least side effects. For now at least, drinking it in moderation is the best policy.

PART IV

Vitamins and Minerals:
Supercharge a Sluggish Metabolism

42

Keep Your Metabolism Fit with Vitamin B

Boost your body's energy production by taking B vitamins. The family of B vitamins plays a critical role in the metabolism of fats, carbohydrates, and proteins. They are also important contributors to cellular energy production. Thiamine (B1) aids the metabolism as carbohydrates are converted into simple sugars. It also helps produce enzymes that are critical in generating energy from food. Riboflavin (B2) plays an important role in the metabolism of protein, fat, and carbohydrates. It is also a vital part of energy-producing reactions at the cellular level and aids the metabolism of toxins in the body.

Niacin (B3) coenzymes are vital to the efficient breakdown of carbohydrates, fats, and proteins. Pyridoxine (B6) is a major player in amino-acid metabolism, the process by which proteins are digested and converted to sugar for energy. It is also intricately involved in the release of sugar from body stores when you need extra energy, such as

Vitamin Number	Chemical Name	RDA for Adults (ages 19 to 50)	Good Food Sources
B1	Thiamine	1.1 mg	Asparagus, sunflower seeds, tuna, peas, beans, tomatoes
B2	Riboflavin	1.3 mg	Liver, mushrooms, spinach, venison, tofu
B3	Niacin	15 to 20 mg	Chicken, tuna, fish, liver, peanuts, lamb, venison, mushrooms, beef
B5	Pantothenic acid	4 to 7 mg	Broccoli, yams, yogurt, mushrooms, sunflower seeds, liver
B6	Pyridoxine	2 to 2.2 mg	Bananas, chicken, baked potatoes with skin, watermelon
B7	Biotin	300 mcg	Egg yolks, liver, leafy greens, cauliflower, sunflower seeds
B9	Folic acid	400 mcg	Leafy greens, asparagus, brussels sprouts, corn, peanuts
B12	Cobalamin	4 to 6 mcg	Shellfish, liver, chicken giblets, egg yolks, lentils, spinach, brussels sprouts

dark leafy greens, peas and beans, dairy products and eggs, fish, poultry and meats, and whole grains. Eating fortified pastas, breads, and cereals is another good way to get B vitamins into your diet. Eight separate B vitamins are required for optimal health, each in varying amounts. The chart opposite shows RDAs and lists the best food sources for each of the B vitamins.

Certain seafood is rich in vitamin B12—particularly clams, which pack a whopping ninety micrograms per serving. Three ounces (85 g) of steamed crab has 8.8 micrograms of B12; three ounces (85 g) of salmon contains 2.4 micrograms; three ounces (85 g) of mussels provide 20.5 micrograms.

Dairy products fall at the other end of the spectrum—an ounce (28 g) of cheese only has 0.5 micrograms. Eight ounces (240 ml) of milk offer nearly one microgram, so the average adult could obtain the daily requirement from two cups of milk and a portion of salmon.

during intense exercise. Pantothenic acid (B5) is necessary for your body to be able to convert fat, carbohydrates, and protein into energy successfully. The B vitamins are water soluble, meaning fluids carry these vitamins throughout your body, and the excess is excreted.

Good Food Sources of B Vitamins

Clearly, the B vitamins play an important role in metabolism function. Most B vitamins can be readily obtained in supplements, but they're also naturally present in a variety of foods. Foods that tend to be high in B vitamins include

It should also be noted there is a general lack of scientific studies supporting the excessive supplementation of these vitamins for enhanced athletic performance. Money spent on B vitamin concoctions may better be spent on the whole foods themselves.

43

Burn Fat with Vitamin C

Would it be better to drink orange juice rather than water while working out? Vitamin C encourages your body to burn fat more efficiently—and faster—as well as speeding up the release of fat from the body. Fat burning becomes even more effective when you combine vitamin C with exercise. Studies show that taking 500 milligrams of vitamin C daily can increase your body's ability to burn fat while exercising by 30 to 40 percent. A 2005 study published in the *Journal of the American College of Nutrition* showed that people can lose weight faster while taking vitamin C, partially due to the way vitamin C helps metabolize fat.

Also called ascorbic acid or L-ascorbate, vitamin C is used in the production of carnitine, a molecule vital in the conversion of fat to usable energy. A 2006 study published in the journal *Nutrition & Metabolism* found that individuals with low levels of vitamin C in their blood oxidized 25 percent less fat per kilogram (2.2 lb) of body weight during exercise than people with normal levels of vitamin C. When a subset of the vitamin-C-depleted group was given 500 milligrams a day of vitamin C, they increased their fat burning during exercise by a factor of four, compared to the rest of the vitamin-C-depleted group.

The test subjects with low vitamin C levels also reported more fatigue while exercising, compared to the control group. It appears, therefore, that by combating fatigue, vitamin C could indirectly contribute to building lean muscle mass, thereby upping metabolic rate.

Generally, too little vitamin C in your diet can lead to a slower metabolism. A 1999 review article in the *American Journal of Clinical Nutrition* stated that vitamin C is responsible for many different aspects of metabolism—indeed, it is necessary for the normal metabolic functioning of the body. It helps with the maintenance of bone and tissue, and, as an antioxidant, aids the removal of toxic materials from the body (▶55).

Vitamin C's Antioxidant Properties Can Raise BMR

BMR (basal metabolic rate) tends to slow down as people age, possibly because the nervous system has an increasingly difficult time keeping that metabolic rate high. A 2003 study published in the *Journal of Clinical Endocrinology & Metabolism* showed

that older adults, between the ages of sixty and seventy-four, who received intravenous injections of vitamin C, had measured increases in their basal metabolic rate of about 100 calories per day. The researchers theorized that the infusion of antioxidants from the vitamin C allowed the bodies of the test subjects to better neutralize free radicals, thus reducing the oxidative stress that can damage cells and tissues and inhibit the smooth functioning of metabolism.

Vitamin C is a water-soluble antioxidant, which means it reduces the damage caused by oxidation within the body.

As your body breaks down and digests food, a type of molecule called a free radical is produced. Free radicals can have damaging effects on cells; antioxidants prevent and reduce this damage.

Many foods are naturally high in vitamin C. These include citrus fruits—oranges, grapefruit, lemons, and limes—as well as vegetables such as broccoli, cabbage, and carrots. An orange contains about 70 milligrams of vitamin C—almost the RDA. Kiwi fruit provides significantly more vitamin C, about 120 milligrams, and guava provides about 180 milli-

grams. An array of supplements is available to augment your diet. Although the RDA for vitamin C is 75 milligrams per day, many experts suggest a daily intake of 400 to 500 milligrams per day to maximize health benefits. More isn't necessarily better, however, and some experts advise staying under 2,000 milligrams per day to avoid possible side effects. Because vitamin C is water-soluble, your body excretes the excess it can't absorb and doesn't need.

44 Generate More Power with Coenzyme Q10

Get more energy to train harder with Coenzyme Q10. This important nutrient helps provide muscles with energy, and if you're deficient in CoQ10 you might not perform efficiently during repeated intense cardio or weight-training sessions. A 2000 study published in the *Journal of Sports Medicine and Physical Fitness* showed that participants who took CoQ10 supplements for an eight-week period were able to sustain higher levels of exertion on an exercise bicycle at the end of the treatment period, although they showed no improvement in aerobic power.

A 1995 study in the *International Journal of Sports Medicine* found no improvement in performance, but did note an increase in the test subjects' reported sense of vigor. Consuming an adequate amount of CoQ10 could theoretically allow you to train harder and build more lean muscle tissue, an absolute necessity to raising metabolism.

CoQ10 Improves Cell Health and Energy Production

CoQ10 plays a major role in metabolism by helping your body convert fats and carbohydrates into a usable form of energy. It's involved with producing ATP (adenosine triphosphate), the primary nucleotide responsible for energy transfer between cells. A coenzyme is a type of molecule that enhances the reactions and abilities of enzymes. Coenzymes usually support a mineral or vitamin, and they can react with a type of protein called an apoenzyme to create an active enzyme. Water-soluble vitamins such as B2 and B6 are examples of common coenzymes. Q10 refers to the fact that this coenzyme is a quinone, or a particular organic compound that helps the body's cells use energy.

CoQ10 functions as an excellent anti-oxidant, or a compound that works to protect other organics from the potential damage of oxygen (▶ 55). Consider what happens when you bite into an apple, and then leave the rest of the fruit on the counter for a few hours. The exposed pulpy fruit turns brown because oxygen reacts with the apple tissues to form o-quinones. CoQ10 fights that same type of damaging reaction when it takes place within the cells of your body. Healthy cells contribute to more effective digestion and less wasted energy. This in turn can result in a more efficient metabolism.

Are You Getting Enough CoQ10 in Your Diet?

CoQ10 is produced naturally by cells in the body, and it's also present in food, particularly animal sources. Young people naturally generate about 300 milligrams of CoQ10 per day, but this amount begins to decrease slowly once you are over age thirty—which might be one reason why we have less energy as we grow older. Most people get less than five milligrams of CoQ10 daily from their diets.

Beef and pork hearts and livers are high in CoQ10, containing ten to twenty milligrams per serving; dairy products and eggs are also good sources. Some types of fish (notably trout and herring) have about one milligram per serving. Leafy green vegetables such as spinach contain around 0.5 milligram per serving, and some citrus fruits have 0.4 milligram per serving. Supplements provide substantially more (usually fifteen to 200 milligrams).

How much CoQ10 should you take daily? Coenzyme Q10 deficiencies are largely theoretical; this means that as far as major medical studies are concerned, no deficiency symptoms have been reported among the general population. For most people, eating a varied diet coupled with your body's natural production of CoQ10

should be adequate. However, if you're looking for a potential boost, start with a low dosage, perhaps thirty to ninety milligrams per day, but consult a health care professional to determine a dosage suitable for your particular regimen. CoQ10 is a fat-soluble supplement, so ideally you should take it with food or milk to optimize its absorption.

This nutrient is found in high concentrations in the heart (as well as the kidneys and other organs), and a 1985 study published in the *Proceedings of the National Academy of Sciences of the United States of America* suggested that heart function might improve with supplementation of CoQ10. Studies have indicated that CoQ10 could also be a factor in preventing certain cardiovascular problems that

might interfere with exercise. CoQ10 may help inhibit the early formation of atherosclerosis by interfering with the deleterious effects of LDL cholesterol.

45

When You Work Out Hard, Supplement with Minerals

After a hard workout, it's a great idea to supplement your system with particular minerals. You lose zinc when you sweat, so if you exercise in high temperatures or for a long period of time you could end up with a zinc deficiency. Zinc regulates cell growth and repair. As discussed in a 2000 article in the *Journal of Nutrition*, this important mineral plays a role in synthesizing protein and DNA. Zinc is known to be an integral part of a strong immune system, which helps your body to function at optimal efficiency.

High-Protein Foods Provide Plenty of Zinc

Zinc is not produced in the body, so it must be obtained from outside sources. High-protein foods tend to be high in zinc, which is good news for metabolism boosters. Meat and fish (especially oysters, beef, liver, pork, and lamb), legumes (beans and peas), whole grains (wheat, oats, brown rice), nuts (cashews, pecans, seeds, peanuts), and dairy products are good sources of zinc.

The U.S. recommended daily allowance (RDA) for zinc is eight to eleven milligrams for adults, more for pregnant and nursing women. More doesn't necessarily mean better, however. Taking exceptionally high amounts of zinc can lead to a copper deficiency (▶ 46). The balance of these two minerals should be maintained as much as possible, and a good nutritionist can help you keep these minerals in check.

Chromium Aids Glucose Metabolism

Another mineral, chromium, is one of the most important trace elements for humans. Chromium supports metabolism by regulating the way carbohydrates are used by the body. Studies show that chromium enhances the effects of insulin. Proper levels of insulin are necessary for the body to metabolize both fat and protein effectively.

Dietary augmentation of chromium has been shown to improve blood sugar regulation, according to a 1998 article in the *Journal of the American College of Nutrition*, and maintaining optimum glucose metabolism is essential to weight control.

Chromium may also help make insulin more effective. A 2006 study published in the journal *Molecular Endocrinology* showed that chromium can increase the ability of cells to transport glucose across cell membranes. Other studies, such as a 1998 article published in the *Journal of Nutrition*, have linked chromium to increased stores of glycogen in the muscles, which can improve endurance during exercise.

Chromium's role in increasing the effectiveness of insulin and affecting fat and protein metabolism have led to claims that chromium alone could reduce body fat and increase muscle mass. Studies, as discussed in a 1999 article in the journal *Annual Reviews of Nutrition*, have not completely borne out these claims, however. Nor have claims that chromium supplements can lead to weight loss been substantiated adequately, in general, by scientific studies, according to a 1998 article in the journal *Nutrition Reviews*.

A more recent combined analysis of a number of studies, published in 2003 in the *International Journal of Obesity and Related Metabolic Disorders*, found a slight weight loss—about 2.4 pounds (1.1 kg)—associated with chromium supplementation. However, this small amount may not be enough to affect overall health or metabolism.

High Levels of Exertion Deplete Minerals
Studies, as reported in a 1986 article in the journal *Metabolism*, show that people who eat a diet rich in simple carbohydrates may be more likely to experience a chromium deficiency. In addition, a 2000 study published in the *American Journal of Clinical Nutrition* showed that high levels of exertion were associated with increased levels of chromium lost through urine. This

study suggests that people who do regular, strenuous exercise may need more chromium than those who are more sedentary. Don't use this as a reason to avoid exercise, however; instead, take it as a reason to increase the *complex carbohydrates* in your diet (▶16).

Chromium is found naturally in a range of foods, including molasses, liver, egg yolks, beef, broccoli, grapes, oranges, and whole grains. Chromium supplements are also available. Recommended levels of chromium consumption are low, minimum values based on assumptions about the average consumption in a typical diet. Studies that have shown the ability of chromium to reduce blood sugar levels used doses of about 200 micrograms per day for a time period of a few months.

46

Consume More Copper to Raise Energy Levels

Increasing your daily intake of copper could help you work out harder and more effectively. A 2005 paper published in the *Journal of Nutrition* found that young men who consumed 1.6 milligrams of copper daily had more energy and were able to exercise at a higher level of exertion than their counterparts who ate diets that only provided the RDA of copper (0.9 milligrams per day). Additionally, the group who consumed less copper had a much higher heart rate when performing moderate exercise.

Healthy Muscles Require Regular Supplies of Copper

Many enzymes require copper in order to function properly. One copper-dependent enzyme enables the cells in the body to produce energy, as discussed in a 1998 paper in the *American Journal of Clinical Nutrition*. This enzyme helps provide the conditions necessary for the mitochondria in cells to create adenosine triphosphate (ATP), which the body needs in order to store energy at the cellular level and build muscle tissue. In the study mentioned above, the activity of copper-dependent enzymes in the muscles of the men who only ate the RDA amount of copper decreased substantially, as compared to those who ingested more copper.

A natural element, copper is essential for the growth of bone, tissue, and most organs. The body also needs copper to properly utilize iron, create healthy red blood cells, and regulate the body's supply of blood and oxygen. In addition, copper aids an enzyme that keeps the connective tissues in the body strong and flexible, so you can exercise effectively. The same enzyme assists in the formation of strong bones and healthy tissues in the heart and blood vessels.

Many studies have shown that a diet high in antioxidants (▶55) can contribute to a speedy metabolism. Copper can act as an antioxidant, reducing the damage caused by oxidation within the body.

A 1996 paper published in the *American Journal of Clinical Nutrition* describes copper's ability to scavenge free radicals, which means that copper can serve as an antioxidant and neutralize compounds that might otherwise break down other cells in the body.

Are You Getting Enough Copper in Your Diet?

Seafood and meat contain copper—one of the best sources is calf's liver. Vegetarians can obtain copper from eating whole grains, legumes, sesame seeds, soybeans, and leafy green vegetables. Certain conditions, such as anemia, may warrant copper supplementation; check with your physician or nutritionist to see if your needs can be met through diet alone.

If you take large amounts of supplemental vitamin C, you should be aware of its possible impact on your level of copper. A 1983 study published in the *American Journal of Clinical Nutrition* found that subjects who took 1,500 milligrams per day of vitamin C for a few months had reduced activity levels in a particular copper-dependent enzyme, although the study participants did not show reduced copper absorption levels.

47 Manage Your Metabolism with Magnesium

Pump up your metabolism and muscle health with magnesium. A 2000 review published in *Clinica Chimica Acta* suggests that magnesium is essential for efficient metabolism, mainly because it helps your muscles to contract and relax properly. Additionally, the chemical reactions in the body that produce energy by burning carbohydrates and fats require magnesium. ATP, the molecule that is synthesized by the body to store cellular energy and fuel metabolic processes, is primarily found in the body in combination with magnesium. Magnesium also is a primary constituent of enzymes that synthesize carbohydrates and fats.

Magnesium Deficiency Slows Metabolism

A 2005 study published in the journal *Diabetes Care* linked magnesium deficiency with obesity and insulin resistance, two conditions that can be associated with a sluggish metabolism. Magnesium helps to regulate levels of blood glucose and prevent diabetes, cardiac disease, and a host of other health conditions.

Magnesium also enables the body to absorb and utilize calcium—the two minerals work together synergistically—which can in turn facilitate weight loss (▶ 23). Like calcium, a large proportion of the body's magnesium stores (approximately half) are located in your bones. Calcium isn't fully absorbed when taken by itself; therefore, magnesium supplements are frequently combined with calcium in order to provide the appropriate balance between these two powerhouse nutrients.

Good Sources of Magnesium

The U.S. recommended daily allowance (RDA) is between 320 and 420 milligrams for adults. Magnesium is found in seeds and nuts, green vegetables, and legumes. Fish (especially rockfish and halibut) and shellfish are also good sources, as are whole-grain cereals and breads. If you are taking zinc supplements (▶ 45), be aware that as little as 142 milligrams of zinc per day can disrupt the body's magnesium balance, according to a 1994 study in the *Journal of the American College of Nutrition*.

Can't stand to take more pills? Try bathing in Epsom salts. They contain magnesium sulfate, which is absorbed by the skin, though it can be difficult to calculate how much magnesium you're ingesting. Ask your physician for advice.

121

48

Inhibit Appetite and Lose Weight with Leptin

Everyone has something called an "appetite center," the part of the brain that tells you when you've had enough to eat. Without the hormone leptin to send these "sated" signals to your brain, you might not know when to stop eating, and may keep on eating even after you're full. Consequently, weight gain and depressed metabolism can result when the body develops "leptin resistance," or the inability to successfully respond to leptin signals.

Studies reported in a 2006 article in the journal *Nature Clinical Practice Endocrinology & Metabolism* have shown that leptin resistance can lead to decreased energy and increased body mass. Other studies, according to a 2008 article published in the *American Journal of Physiology: Regulatory,* *Integrative, and Comparative Physiology*, suggest that leptin resistance leads to weight gain.

Leptin Levels and Body Fat

Secreted by fat (adipose) tissue, leptin is extremely important for regulating energy and appetite, two of the major factors that influence metabolism. Chronic overeating can lead to increased fat stores and excess body weight, throwing your body's delicate metabolic balance out of whack. Leptin levels in the body tend to correlate to your amount of body fat. The more body fat you have, the higher your leptin levels generally are.

Maintaining appropriate levels of leptin may therefore help you to lose weight and elevate metabolism. A 2002 study published in the journal *Circulation* showed that eating a diet high in fish can help lower levels of leptin despite your amount of body fat. For an obese person with suspected leptin resistance, this study gives hope that fish oil may help balance leptin levels. If you eat fish every day, choose low-mercury types such as catfish, herring, trout, fresh salmon, and tilapia. Hate fish? Fish oil supplements are readily available.

Leptin Resistance Impairs Dieting

Some studies, according to a 2000 article published in the journal *Annals of Human Genetics*, have suggested that some people actually have a mutated leptin gene. This correlates with obesity and difficulty losing weight even on a reduced-calorie diet. In cases of genetic leptin deficiency, supplementation

with human leptin has proven to be a successful treatment. However, for people without this mutated gene, supplementing with leptin may actually increase leptin resistance, which can be detrimental for controlling appetite and regulating metabolism.

Numerous name-brand diets espouse the virtues of a "Leptin Diet"; the main principles of all of these diets are fairly similar: Avoid eating late at night, don't consume large meals, load up on protein regularly, and stick to low-GI (glycemic index) carbohydrates and whole grains as much as possible. Although these principles are all sound ways of encouraging the body to burn fat and regulate insulin, ongoing leptin research will prove how effective such methods are at remedying a significant leptin imbalance.

If you are habitually obese and have a low metabolism that doesn't seem to respond to other changes in diet and exercise, you may want to ask your physician about leptin resistance. For people who suspect they have leptin resistance, the dietary changes mentioned above may encourage the body to correct a leptin imbalance, enabling you to control your food intake enough to lose weight.

49

Burn Fat and Rev Up Your Energy with L-Carnitine

To burn fat as efficiently as possible, you need an essential nutrient known as L-carnitine. L-carnitine allows the body to utilize triglycerides (a common type of fat found both in the human body and in the food you eat) for fuel. A 1997 study published in the journal *Progress in Cardiovascular Disease* found that patients who took two to three grams per day of L-carnitine had reduced levels of triglycerides. By burning more fat, you can lose weight, boost your metabolism, and exercise with improved endurance and intensity.

L-Carnitine's Role in Metabolizing Fatty Acids

One of the most important benefits L-carnitine offers is increased energy. L-carnitine is a compound used to transfer long-chain fatty acids across cell membranes into the mitochondria—the powerhouses of a cell—where they are metabolized to help the body make energy. Without adequate amounts of L-carnitine, it is much harder for these fatty acids to metabolize efficiently.

A micronutrient derivative of the amino acid lysine, L-carnitine is produced by your liver and kidneys, but not always at the level for optimal metabolic performance. Several studies have examined L-carnitine's role in exercise performance, with mixed reviews.

A 1985 and a 1990 study, published in the *European Journal of Applied Physiology and Occupational Physiology*, found that in trained athletes L-carnitine supplements had small positive effects on athletic performance. A 1987 article in the journal *Physiologie* found L-carnitine had positive effects on strength and endurance when taken as a three-week supplement and as a one-time dose prior to weight lifting. However, as reported in a 2000 review of L-carnitine supplements and exercise in the *American Journal of Clinical Nutrition*, most studies showed little if any improvement in exercise performance; these studies cited the need for further research on the subject.

L-carnitine may also help reverse age-related decreases in metabolic rate. Recent studies on laboratory rats, as discussed in a 2005 article published in the journal *Clinical Nutrition*, have shown that supplementation with L-carnitine and alpha lipoic acid helps restore the energy-producing functions of aged mitochondrial enzymes, bringing them back to normal levels. These studies, while promising, have not yet been tested on humans. L-carnitine also has antioxidant properties (▶55). Much like vitamin C or CoQ10, L-carnitine can help keep cells healthy (▶43, 44).

Good Sources of L-Carnitine

How much L-carnitine should you consume daily? L-carnitine is a non-essential micronutrient, and has no recommended daily allowance (RDA). Average suggested daily doses generally range from 500 milligrams to 1,000 milligrams per day; as with any supplement, consult a physician before adding supplements to your diet.

L-carnitine can be found in many foods, usually those that are high in protein. Good sources include beef (around 100 milligrams in a four-ounce [113 g] serving), pork (around thirty milligrams in a four-ounce [113 g] serving), and other types of red meat. Dairy products, fish, poultry, tempeh, and legumes are also natural sources of L-carnitine. In general, the body is able to produce or consume sufficient quantities of L-carnitine to sustain adequate levels.

The average person consumes between twenty and two hundred milligrams of L-carnitine per day, but vegetarians or vegans who do not consume any animal products may have daily intakes of less than one milligram per day.

125

50

Corral Fat and Cholesterol with Turmeric

Are you a fan of Indian cuisine? If so, you're already familiar with turmeric's fantastic flavor, but you may not know that this orange-colored spice offers metabolism-boosting benefits. First and perhaps foremost, curcumin (the most active compound in turmeric) is a powerful antioxidant. Studies, as discussed in a 1985 article published in the journal *Chemistry & Pharmaceutical Bulletin*, show that its antioxidant capabilities resemble those of vitamin C (▶ **43**) or vitamin E, although it may not be as readily absorbed into the body. Antioxidants fight the damaging effects of free radicals on cells (▶ **55**); they may also help prevent damage to mitochondria, the cellular powerhouses that help maintain proper fat metabolism. Another study shows a link between turmeric consumption in mice with a lowered insulin resistance, and a decreased risk of developing Type II diabetes. These combined factors allow your metabolism to operate efficiently.

It's no coincidence that many of the foods touted for their metabolic benefits are also useful for lowering cholesterol levels. Studies, as reported in a 1999 article in the journal *Atherosclerosis*, show that turmeric may help lower both cholesterol and triglyceride levels—at least in rabbits—by increasing the amount of cholesterol that is converted into bile in the liver.

Turmeric Can Be an Effective Anti-inflammatory

Turmeric might ease stiffness in joints, too, enabling you to work out with less discomfort and greater flexibility. That's good news if you've had to limit or curtail your exercise regimen because of joint pain.

Ancient Indian and Chinese medicine has long touted this native Asian plant as an anti-inflammatory. For this reason, it is frequently used in these countries to treat swelling and stiffness in joints, as described in a 1972 paper in the *Indian Journal of Medical Research*. Other preliminary studies, such as one conducted in 2006 and published in the journal *Arthritis and Rheumatism*, suggest turmeric may be beneficial for some people who suffer from rheumatoid arthritis. Curcumin's anti-inflammatory properties are well documented in the lab and in animal studies, though larger confirmatory studies are pending.

If breathing problems due to asthma or allergies make it difficult for you to exercise as often or as hard as you'd like, turmeric might offer relief. Curcumin has been postulated to ease some symptoms of asthma due to its anti-inflammatory properties, as discussed in a 2008 study published in the journal *Biochemical and Biophysical Research Communications*. Despite promising research on turmeric's benefits, asthma is a serious disease that should always be treated by a medical professional.

Incorporating Tasty Turmeric in Your Diet

Turmeric belongs to the same plant family (the Zingiberaceae family) as ginger (▶**20**). A staple of Indian cuisine, turmeric is most often found as a bright yellowish-orange powder, though it can also be infused as an oil. A common ingredient in cheese, margarine, and salad dressings, turmeric is frequently used in curries. Add it to meat, vegetables, or fish dishes to gain its many benefits—be careful handling it, though, as it can stain hands and clothing. If the taste is not to your liking, turmeric is available in supplement form.

51 Trim Your Tummy with CLA

For many people—men in particular—abdominal fat is often the last to go. According to a 2000 study published in the *Journal of Nutrition*, conjugated linoleic acid (CLA) can reduce body fat mass in overweight patients. CLA works by interfering with an enzyme (lipoprotein lipase) that increases the size of fat cells. In particular, overweight individuals taking CLA tended to lose more abdominal fat as well as fat from around the rest of the body's midsection.

The same study also showed that people taking CLA tended to gain more lean muscle tissue than fat. Lean muscle further contributes to your ability to burn body fat, and increasing your muscle mass is one of the best ways to raise your basal metabolic rate. A 2003 study published in the *International Journal of Obesity* confirmed these results, finding that participants who took CLA after they had lost weight regained more lean body mass and thus increased their resting metabolic rate.

CLA Supports Cell and Muscle Health

Another metabolic advantage of CLA lies in its ability to function as an antioxidant (▶55). Much like vitamin C (▶43), CoQ10 (▶44), and other antioxidants, CLA helps to protect cells from oxidation damage. By keeping cells healthy, CLA aids digestion and enables other metabolic activities to work more efficiently. CLA may help to lower cholesterol and regulate insulin production and glucose metabolism as well. Different studies have demonstrated mixed results, and more research is on the way to providing clarity.

Get More CLA from Grass-Fed Animals

CLA is a type of fatty acid—it includes a family of different linoleic acids that are found mainly in meat and dairy products. Made by bacteria in the rumen (the fermentation vat in cows, sheep, and goats), CLA can be obtained from ruminant animal sources. If you need an excuse to dine on grass-fed beef, look no further—grazing cattle produce meat with much higher quantities of CLA (three to five times as much) than grain-fed animals. Standard milk contains around four milligrams of CLA per fat gram, whereas milk that comes from grass-fed cows has much more CLA.

Although CLA does not have a recommended daily allowance (RDA), some studies suggest that people may start seeing results by ingesting 3.4 grams

daily. On average, humans consume between fifty and 150 milligrams of CLA per day, which doesn't come close to the 3,000 to 4,000 milligrams believed to aid fat-burning.

Can you get too much of a good thing? Some studies, as cited in a 2004 article published in the *American Journal of Clinical Nutrition*, show that when people who are overweight consume excessive amounts of CLA, insulin resistance and the risk of cardiac disease can actually be increased. Very high doses of CLA may cause upset stomach and diarrhea, but there seem to be few other short-term side effects. As with most supplements and medications, pregnant and nursing women in particular should consult a physician before taking CLA. If CLA is an appropriate supplement, combine it with your diet and exercise plan to start realizing a speedier metabolism.

Get More Energy for Better Workouts with NADH

More energy means the potential for more intense and prolonged workouts, which will burn more calories and help build more lean muscle. In turn this will help raise your resting metabolism. Nicotinamide adenine dinucleotide (NADH), a natural coenzyme that is derived from niacin (vitamin B3, or nicotinic acid, ▶ 42), is essential to the production of energy in body cells. It's a vital cog in the production of ATP, the storage form of usable cellular energy, and you cannot live without the energy provided by ATP. NADH is therefore critical to the functioning of metabolism.

NADH Energizes Your Cells

NADH is a coenzyme, a type of molecule that assists enzymes. Also known as Coenzyme 1 or Co-E1, NADH helps to invigorate your body's cells by increasing the supply of energy. Niacin can help decrease fatty acids in the blood, and is vital when it comes to converting food into energy.

According to a 1998 study by the Nicholas Institute of Sports Medicine and Athletic Trauma in New York, NADH may enhance endurance as well. The study found that triathletes who took NADH had increased endurance and shortened recovery times after competition. Some evidence indicates that in certain populations, NADH can improve overall energy. A 1999 study published in the *Annals of Allergy, Asthma, and Immunology*, and a 2004 article in the *Puerto Rican Health Sciences Journal* suggested NADH supplementation may increase energy in patients who suffer from chronic fatigue syndrome.

As an antioxidant—a substance that works to minimize the damage of free radicals—NADH offers additional metabolism-raising benefits. Free radicals are created during metabolism, and also produced by cigarette smoke, pollution, and other environmental sources. They attack and damage healthy cells; antioxidants like NADH limit damage by neutralizing free radicals (▶ 55).

The body does not produce NADH, so it must be obtained externally. Good sources of NADH include red meat, poultry, fish, and yeast. You can also get NADH in supplement form. Around 2.5 milligrams per day is usually enough for healthy people under the age of fifty; people over age sixty can increase that dosage to as much as ten milligrams per day to keep their energy levels high.

53

Keep Your Thyroid and Your Metabolism Healthy with Iodine

If you're doing everything right, but still can't lose weight, your thyroid could be out of balance. Hypothyroidism, a condition whereby the thyroid gland produces insufficient amounts of thyroid hormones, is typically associated with weight gain, fatigue, and a sluggish metabolism. The thyroid, a butterfly-shaped gland located in your throat, produces some of the most important hormones for metabolism regulation, including triiodothyronine (T3) and thyroxine (T4). Iodine, an essential mineral found in the ocean, plays a key role in the body's creation of these hormones. Iodine is so important, in fact, that we can't function without it.

Insufficient amounts of iodine can lead to hypothyroidism. According to a 2003 article in the *Annals of Nutrition &* *Metabolism*, studies show that a significant percentage of people who eat a vegan diet may be low in iodine. About seven times as many women as men experience hypothyroidism; pregnancy and menopause, and the hormonal changes that accompany them, can affect thyroid imbalances.

Global Concern about Widespread Iodine Deficiency

Iodine deficiency disorders are increasing worldwide, according to the publication *Assessment of Iodine Deficiency Disorders and Monitoring Their Elimination* by the United Nations Children's Fund (UNICEF), the International Council for the Control of Iodine Deficiency Disorders (ICCIDD), and the World Health Organization (WHO). Tens of millions of Americans—

perhaps as many as 60 million, writes Mary J. Shomon in her *New York Times*–bestselling book *The Thyroid Diet*—suffer from this condition, many of them undiagnosed.

Does Your Diet Provide Enough Iodine?

One of the most readily available sources of iodine is salt. Many types of commercially prepared table salt have iodine added to them, due largely to governmental efforts to help prevent a serious thyroid-related disease called endemic goiter, which was common before the twentieth century.

Iodized table salt contains 400 micrograms of iodine per teaspoon, roughly three times the U.S. recommended daily allowance (RDA) of 150 micrograms for adults. During pregnancy your need

for iodine goes up to 220 micrograms, and nursing mothers require about 290 micrograms daily.

However, too much salt in your diet can lead to other health problems, including fluid retention and swollen joints. Fortunately, there are several good non-salt sources of iodine. Iodine occurs naturally in the sea and is found in high concentrations in marine life. Seaweed (kelp, nori, and other varieties) is one of the best food sources of iodine. Saltwater fish and shellfish are also good sources. A three-ounce (85 g) serving of haddock, for example, provides about 120 micrograms of iodine. Shrimp and lobster contain about 25 micrograms of iodine per three-ounce (85 g) serving. Cow's milk, yogurt, cheese, and eggs can also contain iodine, as a result of the iodophors (iodine-containing sanitizing agents) used in making and packaging dairy products.

People on an extremely low-sodium diet should discuss supplemental iodine with their physicians. (Testing can be done to measure your iodine levels and help determine whether supplementation is necessary.) A 2006 article published in the *New England Journal of Medicine* notes that too much iodine can cause thyroid hormone imbalances in people who are susceptible.

54 Increase Energy and Enthusiasm with Ginseng

If you're becoming weary of daily work-outs, diets, and the discipline needed to keep up your regimen, ginseng could spur your energy as well as your enthusiasm. This edible root has long been touted for its ability to ward off mental and physical fatigue. People who exercise at a high level of intensity may benefit most from ingesting ginseng. According to Russian doctor Israel Breckhman, "Ginseng stimulates both physical and mental activity and strengthens and protects the human organism when undergoing severe and/or physical strain."

Ginseng Can Enhance Strength and Stamina

Ginseng (a member of the *Panax* genus) has been used for thousands of years in the East to increase stamina, calm stress, and promote alertness. The first book on herbal pharmacology, published two millennia ago by the naturalist Sheng Neng Pen-T'sao, discusses the energy-boosting benefits of ginseng.

A more recent study conducted at the Department of Exercise Science at Chungbuk National University in Korea and published in the *Journal of Sports Medicine and Physical Fitness* in 2005, supports this claim. The study found that *Panax* ginseng extract "significantly increased exercise duration until exhaustion" in young men, and also facilitated the men's recovery after they had engaged in exhaustive exercise.

Ginseng's ability to enhance strength and stamina had been demonstrated previously in animal studies, conducted at the Department of Physiology in the University of Leon in Spain and reported in the *Brazilian Journal of Medical and Biological Research* in 2004. The study found that ginseng extract protected the muscles of acutely exercised rats by reducing stress and oxidation.

Body and Mind Respond to Ginseng

Exercising purposefully every day requires sustained effort, which ginseng may be able to support. The brain derives its energy from glucose, and ginseng may help the brain increase its uptake of this energy source.

Studies suggest ginseng has a positive effect on mood and cognitive performance. A 1999 study published in the *International Journal of Clinical Pharmacology Research* shows that people who regularly take ginseng supplements tend to have improved outlooks and are more optimistic than they were before taking the supplements.

Ginseng could also help you lose weight. Studies show it can lower cholesterol and triglycerides, the excess calories your body stores in your fat cells when you eat more than you need for fuel. A study performed at the Department of Health and Kinesiology of Purdue University, Indiana, and reported in *Pharmacological Research* in 2003, showed that ginseng assists with the metabolism of carbohydrates and enhances the body's ability to transport and dissolve fats and cholesterol around the bloodstream.

Ginseng is native to eastern Asia (Korea, Siberia, and China). The root of the ginseng plant contains chemicals called ginsenosides, the elements thought to be responsible for ginseng's power. The dried root is formulated into pills, capsules, teas, extracts, and even lotions. A dose of 200 milligrams per day has been proven safe, but always get advice from your physician, especially if you are pregnant, nursing, taking blood thinners, or have diabetes or heart disease.

55

Prevent Cell Damage with Antioxidants

Does the mythological fountain of youth really exist? Perhaps not, but antioxidants may be the next best thing. Experts suggest that antioxidants can slow the aging process by neutralizing free radicals and their potentially damaging effect on cells.

Studies, as discussed in a 2000 article published in the journal *Science*, show that antioxidants can help prevent the damage that occurs to mitochondria during the aging process. One of several kinds of specialized structures within cells, mitochondria are often referred to as "cellular powerhouses" because they convert nutrients into energy. An article published in 2000 in the journal *Free Radical Biology and Medicine* reported on the design of antioxidants meant to target mitochondria directly.

Antioxidants Deter Muscle Degeneration and Disease

Antioxidants are chemical substances that act to reduce the damaging effects of oxidation, or the way in which oxygen interacts with everything around it. Oxygen is critical to human survival in that all cells in the body require oxygen, but oxygen is a double-edged sword. If you've ever seen rust on an iron dumb-bell or a bitten-into piece of fruit turned brown, you've witnessed the downside of oxygen. Inside the human body, oxidation can be nasty.

Whenever the body breaks down food for digestion, unstable compounds called free radicals are produced. These free radicals destroy cells by causing them to become unstable; serious disease can result from the deleterious effects of free radicals. Antioxidants mop up free radicals and remove them before they can cause cellular damage, thereby reducing potential muscle, organ, and tissue degeneration. Maintaining consistently high levels of antioxidants in the body can also promote your immune system's protective response to foreign toxins such as cigarette smoke and environmental pollution.

Keep the cells in your body healthy by including lots of antioxidants in your diet. Fruits, veggies, and whole grains possess antioxidant qualities. Whenever possible, choose whole foods first—you'll also gain the fiber, nutrients, and other important benefits that whole foods provide. When it comes to fruit, think red: red grapes, tomatoes, pomegranates, cherries, and berries offer high amounts of antioxidants. Orange vegetables like carrots are also excellent sources because they contain beta-carotene, a powerful antioxidant. Leafy green vegetables (such as spinach) contain high amounts of lutein, an antioxidant that can also help prevent vision problems. Cruciferous vegetables, or ones belonging to the cabbage family, contain an antioxidant called indole 3-carbinol; broccoli, kale, brussels sprouts, and cauliflower provide this potent antioxidant.

Drink to Your Health: Antioxidant Beverages

If you prefer to drink to your health, try tea: both green (▶**39**) and black teas provide high amounts of antioxidants. They also contain caffeine, a stimulant that can increase your energy and metabolism (▶**40**) but might not be desirable right before bed or if you're pregnant or nursing. Most teas are available in decaffeinated varieties that still retain their antioxidant properties.

Pomegranate juice is also becoming increasingly popular as a health tonic. A 2005 paper in the *Proceedings of the National Academy of Sciences* show that pomegranate's antioxidant properties can be beneficial for fighting heart disease, in addition to promoting a healthy metabolism and immune system. Blueberry juice, or pomegranate-and-blueberry juice, offers similar health advantages with a less intense flavor.

You'll find a huge number of antioxidant supplements on the market as well, including vitamin C (▶**43**), vitamin E, green-tea extract, and beta-carotene supplements. Antioxidant minerals include selenium, zinc (▶**45**), and calcium (▶**23**).

Most stores also carry generic "antioxidant blends" that are a combination of vitamins and food extracts. Generally, though, getting antioxidants through whole foods is preferable to taking a pill. Reserve supplement usage for either nutrients that aren't readily found in foods or for those times when you need more than you can reasonably consume in a day's worth of healthy meals.

56

After a Hard Workout, Recover with D-Ribose

You have to expect a degree of stiffness and sore muscles after a vigorous workout, but D-ribose, a naturally occurring organic compound, can reduce those reactions and help you improve your performance level. D-ribose enters the picture during the recovery phase, as it helps replenish the body's supply of ATP, or adenosine triphosphate, often called the "energy of life."

A 2004 study published in the *American Journal of Physiology* measured the muscle ATP levels of subjects who performed vigorous exercise, and then took either D-ribose or a placebo. Muscle ATP levels dropped after exercise for both groups, but the group that took D-ribose had a lower drop. Furthermore, seventy-two hours later, the D-ribose group had higher ATP levels than the placebo takers, although there was no significant difference in exercise performance between the two groups.

D-Ribose Keeps Your Energy Level High

During intense exercise, levels of readily available ATP within the body fall as energy stores are depleted. This drop in ATP can be associated with feeling tired but also with muscle fatigue, stiffness, soreness, and cramping. D-ribose is a monosaccharide that is required for ATP production. Aside from being a required element of RNA (ribonucleic acid, which, along with DNA, forms the blueprints of human cells), D-ribose promotes metabolism by keeping the body fueled with energy. One of the ways it works is by its association with ATP, a nucleotide whose primary role is to transfer energy within individual cells during metabolism.

D-ribose tends to be particularly beneficial for certain groups of people. Studies such as the one described above suggest that D-ribose may help serious athletes, or anyone who exercises to exhaustion several times a week, to recover faster than they would without adequate levels of D-ribose.

Similar studies, such as one published in 2006 in the *Journal of Alternative and Complementary Medicine*, indicate that D-ribose may be effective at raising energy levels for people with chronic fatigue syndrome and fibromyalgia. Other studies, such as one published in 2003 in the *European Journal of Heart Failure*, have demonstrated how supplementing with D-ribose can be beneficial for certain types of cardiovascular disease. By aiding these conditions, D-ribose may enable those affected to increase the amount of exercise they get.

If you undertake intense workouts without recovering fully each time, you're much less likely to continue exercising on a regular basis. Although the body will eventually heal, the mind can be much less forgiving. Maintaining adequate amounts of D-ribose might enable you to recover more quickly, so you'll be more likely to continue exercising every day. And daily exercise is one of the best ways to raise metabolism, as it helps burn calories and fat while building lean muscle tissue.

Can D-Ribose Supplements Help You?

D-ribose can be obtained from brewer's yeast and meat, but it is difficult to consume much because most of what is contained in those products is eliminated during the cooking process. Supplements of D-ribose come in a variety of forms, including powders and capsules. The amount you should take depends greatly on your physical health, and how much supplementation your individual body requires. Dosages of five to twenty grams per day are not unusual.

D-ribose supplements often start taking effect in as little as a few days. Suddenly ceasing to take D-ribose supplements, however, can result in a loss of energy within a few days. If you're considering taking a new supplement, consult with your physician to determine a dosage.

57 Metabolize Glucose, Fat, and Protein Effectively with Vanadium

Maintaining a harmonious relationship between blood glucose and insulin production is essential to a healthy metabolism and weight maintenance. According to a 1998 review article in the *Journal of the American College of Nutrition*, the chemical element vanadium may aid this relationship. The human body burns glucose for energy; insulin, created by the pancreas, is required to metabolize glucose. When the body fails to efficiently utilize glucose for energy, an excess of blood glucose can lead to diabetes. Vanadium has many insulin-like characteristics that affect not only glucose metabolism, but also fat and protein metabolism as well.

As the *Nutritional Supplement Review* points out, vanadium's insulin-like effects might "enhance endurance and performance by increasing transport of glucose into muscle and fat cells, which may ultimately have some muscle-building effects." For this reason, bodybuilders and others who wanted to increase muscle mass—or at least make their muscles look bigger and fuller—have found vanadium attractive.

Vanadium has also been thought to help increase the ability to exercise at high intensities. However, when vanadyl sulfate, a supplement derived from elemental vanadium, was tested in a controlled study of weight-training athletes, only modest enhancements in performance were found in the study group that took the vanadyl sulfate. According to a 1996 article in the *International Journal of Sport Nutrition and Exercise Metabolism*, no changes in body composition were found. More studies may be necessary to adequately assess vanadium's value in these areas.

Vanadium's Role in Regulating Blood Glucose Levels

Several studies, including one cited in the *Journal of Clinical Endocrinology & Metabolism*, show that vanadium may be helpful in regulating blood glucose levels. Better glucose metabolism can lead to improved energy production and may aid weight loss. Under normal circumstances, glucose penetrates the body's cells, and blood glucose levels fall accordingly. Often diabetes and obesity are linked. For someone with diabetes, the pancreas either fails to produce sufficient insulin or the body's cells become resistant to the effects of insulin; this results in too much blood glucose. For diabetics, in particular, exercising regularly and maintaining an appropriate weight are essential—and vanadium may help. (However, diabetes is a serious condition and treatment should always be handled by a qualified medical professional.)

Raising and maintaining your metabolic rate goes hand in hand with regulating cholesterol levels and blood pressure. Animal studies, such as a 1994 study published in the journal *Hypertension*, show that vanadium may also be able to lower and regulate blood pressure. Other animal studies, cited in a 1990 article from the *Journal of Environmental Pathology, Toxicology, and Oncology*, show that vanadium helps block the buildup of cholesterol. Cholesterol is a fat-like substance found normally in cells

and the bloodstream. Obesity, eating a diet high in saturated fat, and/or not getting enough exercise—behaviors that can hamper your metabolism—can contribute to high levels of LDL cholesterol, the so-called bad cholesterol, which can in turn lead to a range of cardiac problems.

How Much Vanadium Do You Need?

How much vanadium we need is still a subject of debate, largely due to lack of human studies on the subject. We know that vanadium is required for normal development; exactly how much is required has not been scientifically determined. The average daily intake of vanadium is less than fifty micrograms per day. Typical vanadium doses range from ten to sixty micrograms per day; larger doses can produce unwanted

side effects, including high blood pressure and damage to red blood cells. Vanadium supplementation isn't advised for people with diabetes or hypoglycemia, unless recommended by a medical professional.

Your best bet may be to incorporate vanadium-rich food sources into your diet wherever possible. Olive, corn, safflower, and sunflower oils all contain vanadium. It is also found in vegetables (especially mushrooms, carrots, cabbage, green beans, and radishes), herbs (parsley and dill in particular), black pepper, grains (including oats, rice, and buckwheat), and some fortified cereals as well. Vanadium is also present in liver, eggs, herring, and some shellfish.

58

Turn Back the Clock with Human Growth Hormone

If you're inching toward middle age and find you have to work out harder to stay in shape, it could be due to a decrease in your body's levels of human growth hormone (HGH). According to a 1990 article in the *New England Journal of Medicine* that focused on HGH supplementation, HGH levels start diminishing after the age of forty, resulting in reduced muscle mass and increased fat levels. Also known as somatotropin, HGH is produced by the pituitary gland and is responsible for stimulating cellular growth and reproduction. Therefore, HGH can have a powerful impact on raising metabolism.

HGH Helps Burn Fat for Energy

HGH also improves fat burning. The body's first choice of energy, especially when you're working out, is glucose. The pancreas releases insulin that facilitates the movement of glucose into the cells to be used for energy. Extra glucose that isn't needed immediately is converted into fat and stored. Research presented in a 1990 *New England Journal of Medicine* article shows that HGH can help prevent the body from converting as much surplus glucose to fat. As a result, the body uses a higher amount of fat for energy. The mechanism by which HGH affects fat stores stems from its production of IGF-1, an insulin-like growth hormone that is excreted from the liver. HGH can increase the amount of IGF-1 produced, which helps the body accelerate the rate at which it burns fat.

A chapter in the 1997 book *Anti-Aging Medical Therapeutics* confirms that HGH can help increase your energy level during a workout. If you have more energy, you are able to exercise at a higher intensity for a longer duration, enabling you to burn more calories and build muscle tissues.

Exercise and Diet Can Step Up HGH Production

Studies suggest that the HGH produced during anaerobic exercise may be even more helpful for losing weight and regulating metabolism than the synthetic version. You may be able to increase HGH levels simply by doing strenuous anaerobic activity, like strength training (▶2) and interval sprints (▶8).

One of the best ways to raise your metabolism is by strengthening muscles during interval training. An article published in the *Journal of Sports*

Sciences in 2002 confirmed that exercise can cause a rise in the body's levels of HGH. The article also demonstrated that doing sprint work (short bursts of fast, hard activity) can increase HGH production more than other types of exercise. Working out harder and at a higher intensity can lead to higher production levels of HGH, which can translate to more lean muscle tissue and a speedier metabolism.

Reducing high-sugar, high-GI carbo-hydrate consumption is another way to increase the body's natural production of HGH. Studies, as cited in a 2001 article published in the *American Journal of the Medical Sciences*, show that the body tends to produce less HGH when there are higher levels of insulin circulating in the body—and eating foods with a high glycemic index value increases the body's production of insulin. Choosing complex carbohydrates (▶16) instead of simple ones (for example, whole-grain bread instead of white, or a sweet potato instead of a white potato) can decrease the body's insulin response, which may result in a more normalized production of HGH.

HGH Supplements Entail Risks and Costs
Supplementation with HGH can help reverse the age-related decline of this necessary hormone, especially

in people with a diagnosed growth hormone deficiency. However, HGH supplementation can be prohibitively expensive, and there is little compelling evidence that adults who don't have a diagnosed growth hormone deficiency will reap benefits from supplements. Taking synthetic human growth

hormone must be supervised by a physician, and safety concerns may exist for some individuals. Also, be aware that synthetic human growth hormones are banned by many professional sports. The best way to naturally boost your HGH levels is to exercise on a regular basis and stick to healthy foods.

PART V

Lifestyle Changes: Quick Steps
toward a Faster Metabolism

59

Set Goals for Metabolic Success

Raising metabolism and losing weight involve more than physics and mechanics. Being mentally, spiritually, and emotionally committed to improving your health and lifestyle is the only way to ensure that those changes will succeed in the long term. To provide a yardstick by which to measure your progress, create clear, obtainable goals. Studies, as published in a 2002 article in the *Journal of Instructional Psychology*, confirm that people who set goals tend to have higher rates of achievement. Set yourself up for success.

Establish Specific, Achievable Fitness Goals

First, define your goal clearly. Write it down in a goal book, diary, or journal. Putting that goal in writing makes it tangible, and also begins the important process of taking responsibility for meeting your goal. Participants in a study of overweight men and women conducted by Harvard School of Public Health and Pennington Biomedical Research Center in Louisiana, and published in a 2009 article in the *New England Journal of Medicine*, kept online journals to chart their progress during the course of the study. Many of them believed that keeping a diary made them more aware of what they ate and how it contributed to weight gain.

Be specific when setting goals for yourself: "Raise metabolism" is not nearly as useful as setting a particular weight to get down to by a certain date, for example. And give yourself a deadline; this keeps you motivated and gives you a timeline for meeting your goal. You can even add intermediate deadlines, such as to lose a set amount by a certain day of each month.

Next, add a level of detail to your goal by determining how many calories per day you need to eat in order to lose weight. Use one of the formulas on page 13 to figure out your individual BMR, or the number of calories you're currently requiring at rest every day. Factor in the calories burned during your daily activities. If your goal is to lose one pound (0.45 kg) per week, subtract 500 calories per day (which adds up to 3,500 calories, or one pound [0.45 kg], per week) from your maintenance level.

Writing that amount down in your goal book will help you break your goal into reasonable tasks, as well as confirming that you understand what you need to do. It may also be useful to write out a few days' worth of sample meal plans that fall within your caloric requirements.

Record Your Progress in a Health Journal

The next step is to put your commitment to exercise in writing. Suppose you want to burn off 400 calories a day to meet your weight-loss goal. If you weigh 150 pounds (68 kg), you're going to need to put in close to an hour on a recumbent cycle, forty minutes of swimming laps, or about fifty minutes of racquetball. (You'll burn more calories if you're heavier, fewer if you're lighter.)

Once you've written down your goal statement and means to achieve that goal, double back and reevaluate. Is the goal achievable? Can you really stick to an eating regimen that will let you lose at least one pound (0.45 kg) per week? If you're not prepared to sufficiently decrease your food intake and/or exercise hard enough, adjust your initial goal. Setting yourself up for success gives you confidence to achieve your current goal and enables you to set your subsequent goal higher.

Finally, make an effort every single day to focus on your goal. Many people find it helpful to write in their goal books or journals at least once a day. Note

areas where you've done well ("skipped dessert, worked out an extra twenty minutes on the cycle") and areas where you feel you could have been better ("got the large popcorn at the movies instead of the small box"). Understand that you, and you alone, have the power to achieve your metabolic goals.

60

Count Calories Before Consuming Them

It is easier than you might think to unknowingly overload on calories. How much damage can one tiny truffle do? Plenty. One chocolate truffle contains at least eighty calories, with more than half of those calories coming from fat. One slice of pecan pie can weigh in at 500 calories. One of the first steps in limiting metabolic decline is learning which foods pack on the weight, especially some you might have thought were innocuous.

A study of 811 overweight men and women conducted by Harvard School of Public Health and Pennington Biomedical Research Center in Louisiana, and published in a 2009 article in the *New England Journal of Medicine*, found that what you eat is less important than the number of calories you consume.

According to Dr. Elizabeth Nabel, director of the National Heart, Lung, and Blood Institute, which funded the study, "The hidden secret is it doesn't matter if you focus on low-fat or low-carb."

Learn to Recognize the Calories in Food Portions

Let's start with something simple. What does fifty calories look like? Two cups of air-popped popcorn, a handful (fifteen) of grapes, or three pieces of thin-sliced turkey all have about fifty calories.

How about 100 calories? One egg, three ounces (85 g)—about a palm-sized piece—of white fish, one large apple, a cup of blueberries, a handful (fifteen to twenty) of peanuts, or about forty mini pretzel sticks weigh in at around 100 calories.

Two hundred calories worth of food becomes more substantial, what you might consider a snack. A medium-to-large baked potato has around 200 calories, so if you want to enjoy it with sour cream or a little cheese on top, choose a smaller potato and use a minimal amount of toppings. A seven-ounce (200 ml) cup of chili has about 200 calories, without extra toppings. Four fast-food chicken nuggets (with a little sauce) contain 200 calories, as does a four-ounce (113 g) serving of roasted ham or half a bagel with two tablespoons of cream cheese.

Portion size is obviously important when it comes to counting calories. Two slices of avocado may add up to fifty calories, but if you end up eating the entire fruit on your sandwich, that's 320 calories. One slice of bacon has about forty calories, but eating five slices at breakfast quickly comes to 200 calories. Calorie-dense foods, which often are high in fat as well, can add up quickly. Watch out for things like full-fat dairy products, butter, and salad dressings.

Keep a Close Eye on Your Calorie Consumption

Some people find it helpful to write down everything they eat each day in a food journal (▶61). You record the name of the food, the portion size, and number of calories. (These values can easily be obtained from a calorie-counter book or nutrition web page.) When tracking your food intake, be honest with yourself. If you see that a serving of cashews (fourteen to seventeen cashews) has 165 calories, but you really ate closer to thirty, note your caloric intake as 330, not 165. Fudging the numbers won't help you in the long run, as your metabolism knows the truth.

Just because one item has a certain number of calories, don't assume that a similar item contains the same. A cup of cornflakes or oat-rings cereal has 100 calories, for example, but a cup of granola comes close to 600 calories. A four-ounce (113 g) serving of 95-percent lean ground beef has 150 calories, whereas the same size serving of 70-percent lean ground beef will give you more like 350 calories. Do the math before eating, and choose wisely.

Reading package nutrition information is the best way to gauge how many calories you're getting from packaged foods. Many treats (crackers, cookies, and even some candies) now come in 100-calorie packs, so you can see exactly how much you can eat of your favorite snack. These products almost always cost more than a regular package of the same treat, so once you become familiar with the appropriate portion size, buy full-sized bags and just count out one serving of chips or nuts.

Learning to recognize and remember the caloric content of your favorite foods will take time, but it's well worth the effort. Soon you'll be able to see your daily intake add up, and choosing appropriate foods will become second nature.

61

Keep Tabs on Calories with a Food Journal

Whether you eat three full meals, five small meals throughout the day, or meals plus snacks, make a habit of writing down what you eat. Food journaling promotes a self-awareness that can help you shed weight. A 2008 article in the *American Journal of Preventive Medicine* confirmed that overweight people who logged their meals tended to lose significantly more weight—up to twice as much—as those who did not keep food journals.

Writing It Down Makes You Aware of and Accountable for What You Eat

Keeping a record of what you can eat (and when) helps prevent mindless snacking, or eating for reasons other than hunger. The simple act of recording what you eat makes you aware of all those extra calories you either didn't

realize you were consuming, or didn't want to know about. Writing down *everything* you eat also creates a sense of accountability—and you may find this external form of motivation provides just the boost you need. Pamela Peeke, M.D., author of *Fit to Live,* agrees. "Journals are a form of accountability … that help reveal which strategies are working. When you are accountable, you are less likely to have food disassociations, or be 'asleep at the meal.'"

Choose a spiral- or hard-bound note-book specifically for this purpose, and get in the habit of noting every single thing you eat. Familiarize yourself with the calorie content of the foods you eat (▶ 60), or look them up in a calorie-counting book or website (a good site is www.sparkpeople.com). Can you just as

easily create an online food journal? Of course, but print out some blank pages and leave them in the places your snack foods live: the refrigerator, pantry, cabinets—anywhere you're likely to nibble.

Will this be an embarrassing activity? Possibly, once you learn the cumulative effect of "a little bite here" and "just one sip" there. But instead, let this be an enlightening activity, a wake-up call. The bottom line is: All calories count.

Control Calorie Intake by Planning Metabolism-Boosting Meals

In your food journal, compile a variety of metabolism-boosting meals. Each day's worth of meals should total your daily caloric limit. Numerous books, organizations, and websites provide

losing weight and not toward raising your metabolism. Although the two are not mutually exclusive, a metabolism-boosting diet may not look exactly the same as a weight-loss diet. Use the information in this book as a starting point for creating healthy meal plans, ones designed to raise metabolism. As you become more familiar with the process, you'll get more inventive and expand your repertoire of meals.

Planning meals prevents unintentional overeating (using more cooking oil than you need, for example, or adding a cup of walnuts to a salad when the recipe only calls for a quarter cup). Being aware of everything that goes into your mouth is a giant step toward gaining control over your caloric intake. Creating meal plans also helps you choose the right foods to eat at the right times. A little organization can go a long way toward keeping metabolism high all day long.

Honesty is the best policy when it comes to adhering to a meal plan. If you eat an extra bag of chips with lunch, or finish the chicken nuggets from your child's plate, note it in your journal, so you can adjust your calorie intake for the rest of the day. If you find that you're over your daily count before dinner, try to fit in an extra walk, and make sure your evening meal has fewer calories than usual.

planned meals with a preset number of calories, but there are a couple of reasons not to rush into signing up for any particular program. For starters, many of these plans are expensive. They also may be tailored toward

62

Record Your Workouts in an Exercise Journal

When it seems you barely have time to cram exercise into your schedule, taking a few extra minutes to write down the details of your workout may seem like overkill. However, there are many good reasons to keep an exercise journal—and many gyms provide materials that let you record your progress.

Tracking Your Achievements Keeps You Moving Forward

First, you have to keep an accurate record of your progress in order to know what areas need improvement. Journaling allows you to keep track of how many repetitions you did at each machine, which in turn creates a history of which muscles were trained (and for how long) at your last workout. Can't remember how far you biked last week? Writing it down in a journal commits your achievement to more than just your memory. It gives you a good record of how far you already know you can go, and how much you've tested your body to reach that point.

Second, journaling enables you to set goals. Once you know where you've been, you can see more clearly where you're going. Writing down reps and weights will let you know if you are moving forward or stuck on a plateau, and will give you insight into how you can adjust your workout to continue building muscle and burning fat. When biking your usual distance becomes so easy that you're doing it in a shorter time and without breaking a sweat, you know that it's time to increase the intensity of the workout in order to keep revving up your metabolism.

Underneath these two facets of journaling (keeping track of your past achievements and delineating your goals) lies the real benefit to the metabolism booster: being able to clearly see your progress. While a drop in dress size or noticeably bigger biceps are definitive and undeniable clues that you are losing fat and gaining muscle, those changes don't always happen immediately. Your body could be gaining metabolic speed before you see it reflected in your body, and an apparent lack of progress can be a major mental deterrent to continued daily exercise. Writing down your progress after each workout reminds you of how hard you've worked to raise metabolism and improve health—and that reminder may be all you need to get back into the gym the next day.

Monitor Both Cardio Activity and Weight Training

A typical exercise journal page should be split into two sections: cardiovascular activity (running, walking, cycling, and so on) and weight training. Start each page with the date, and your weight if you're stepping on the scale every morning. If you're not trying to lose weight or if you weigh yourself less frequently, enter it as often as you feel is necessary to track your progress. Weighing yourself once a week will help you avoid stressing about the daily variances of weight, and keep you focused on the means to the end.

The weight-training section should include columns for amount of weight, number of repetitions, number of sets, and the name of the machine or exercise. Also note the parts of your body being worked by that exercise (biceps, triceps, and so on). The cardio section of your journal should contain columns for type of exercise and number of calories burned, if you're using a machine or monitor that provides that information; if not, you can look it up later in a book or online resource.

Make sure to record how long you exercised, and at what intensity. Most machines at a gym have this information readily available—note that you worked out at Level 4 on the elliptical trainer, for example, or Level 5 using the "Cross Country" program. If you're walking outdoors, note the distance you covered and the amount of time you spent walking. Wearing a pedometer lets you record how far you walked in a workout session or during a typical day.

Don't limit your journal to repetitions, miles, and weights. Feel free to also jot down your emotional state before, during, or after a workout. Did you feel good about your workout? If you stayed up too late the night before, were you low on energy? Did you leave the gym feeling tired and sluggish, or energetic and psyched for your next workout? Was last night's dinner still lying heavy as a rock in your belly? Did a particular song you listened to while cycling help pass the time?

Being aware of how you're feeling and what you're experiencing, mentally and physically, can help you see what corrections you need to make in your daily routine, and those corrections will allow you to maximize your workout efficiency and boost your metabolism.

63

Exercise Early to Elevate Your Metabolism All Day

Morning exercise stimulates your metabolism for the entire day. A study presented at the North American Association for the Study of Obesity 2005 Annual Scientific Meeting showed that early-morning exercise had a measurable effect on the blood levels of triglycerides all the way through the day until 10:30 p.m. The lasting effect of a moderate, one-hour, prebreakfast workout shows how exercise can have a long-term impact on the metabolic processes, including fat burning.

Additionally, exercising early in the day helps your natural rhythms reset at bedtime so you can get a restful night's sleep and wake up ready to start the next day with another great workout. A 2003 study published in the journal *Sleep* found that exercise in the morning enabled study participants to sleep better at night. And studies done at Stanford University and the University of Chicago in 2004 showed that sleep deprivation could be linked to obesity (▶ 26, 84).

Morning Exercise Revs Up Calorie-Burning

Every time you get your heart pumping through exercise, you boost your metabolism—and your metabolic rate remains higher for a few hours after exercising. By exercising early in the day, you're treating yourself to a higher metabolism from the outset. You're helping your body to digest food faster and burn more calories in the process. Morning exercise can also help you feel more alert throughout the day. In contrast, exercising late in the evening can stimulate you and keep you awake, instead of shifting you into a calm and relaxed state for sleep. Working out in the evening will still temporarily raise your metabolic rate, but you won't be able to take advantage of that faster calorie-burning mode as efficiently while you're sleeping as you would when you're awake. Metabolism slows down while you're sleeping, so you won't get the full advantage of that metabolic revving if you exercise right before bed.

Exercising early also shows your commitment to making exercise a part of your daily routine. For many people, it's easy to say you'll exercise after work, or after dinner, or during lunch, but the reality is that many of us don't stick to a plan that isn't firmly engrained in our schedules. Eighty percent of people who

exercise daily work out in the mornings. By dedicating an hour each morning to exercising, you're more likely to (a) remember to do it, because it's the first thing you do every day; (b) eliminate excuses for avoiding exercising, as your social and work calendars will not have impinged upon your time yet; and (c) set a positive tone for your day. The mental component of morning exercise cannot be underestimated. By establishing a pattern of prioritizing exercise, you're priming your body and your mind for the day. Morning exercise can also help regulate appetite throughout the day, encouraging you to make better food choices, which can help you reach your weight goals.

Working out in the morning can help your metabolism in a number of ways, but you'll lose many of those benefits if you think you can eat whatever you want during the rest of the day. Burning 200 calories in a cycling session entitles you to an extra 200 calories of food intake—that means two extra pieces of fruit throughout the day, for example— but only if you're trying to maintain your current weight. If you're trying to lose weight, you need to aim for a caloric deficit. That means you still need to stick to the plan: small, nutritious meals spread throughout the day will help keep your metabolism high (▶34).

64 Eat After Working Out to Boost Your Metabolism

Although eating before a workout can provide additional energy, eating after you exercise—while your metabolism is still running high—is even better. If you eat the right foods at the right time you can raise your metabolism more than you would through exercise alone. Studies, such as one cited in the 2001 book *Eat Smart, Play Hard* by Liz Applegate, show that eating properly after exercising can help you work out more intensely the next day as well.

When you exercise, you boost your metabolism, and this metabolic over-drive keeps running for several hours after you finish exercising. Therefore, you'll burn calories more quickly and efficiently if you eat after a workout than at times of the day when your metabolism is naturally running slower,

like right before bed. Of course, you're still consuming calories, but you can take advantage of a faster metabolism by eating within an hour of exercising. Some bodybuilders and high-intensity athletes will down a protein bar or shake immediately following a workout, but for most of us, eating fairly soon after exercising is enough to keep your metabolism charged.

Feast on Protein and Carbohydrates after a Workout

Generally speaking, the two main components of any post-exercise meal should be protein and carbohydrates—ideally ones that are fairly easy on the digestive tract. Professional sports teams have researched the best way to refuel their athletes after intense weight-lifting sessions, and have found that a four to

one ratio of carbohydrates to protein gives lifters a small but measurable increase in strength and recovery. Believe it or not, chocolate milk is one of those post-lifting indulgences that falls within the optimal ratio recommended by some professional trainers.

Exercise depletes your body's reserves of glycogen (the stored form of glucose that provides you with a ready source of energy). Glycogen stores are rebuilt most quickly from carbohydrates; therefore, complex carbohydrates are an excellent choice for part of a post-workout meal (▶16). Good choices include whole-wheat bread, brown rice, fruits, or vegetables. In addition, the body restores glycogen most efficiently during a short period of time—up to about an hour—after exercising, so that's the ideal time for a healthy meal. Can't live without your bagel and orange juice? The best time to consume these types of quick-energy carbohydrates is right after exercising—so go for a jog first.

Protein is a good choice for a post-exercise meal, as it helps to strengthen and repair muscles (▶17). A 2002 study published in the *Journal of Applied Physiology* showed that muscle glycogen reserves increased twofold when protein and carbohydrates were consumed together after exercising. In the study, participants exercised vigorously, and then some were given a supplement containing protein and carbohydrates; others received only carbohydrates. The test subjects' glycogen levels fell immediately post-exercise, but rose faster in participants who took the protein-carbohydrate supplement than in those who took the carbohydrate-only supplement. Choose low-fat protein sources, such as lean meat or low-fat dairy products. Protein shakes let you ingest high amounts of protein quickly, but beware of the additional sugars and calories many of these mixtures contain.

Eat Light After Exercise to Optimize Fat Burning

Avoid eating large quantities of fat after exercising. Fat takes longer to digest than carbohydrates, and will lengthen the amount of time before you'll be able to use the healthy carbohydrates in your meal for energy. Steer clear of high-fat proteins such as whole milk cheese, peanut butter, or fatty red meats. A heavy meal can make you feel tired at a time when you should be feeling energized. Eating a small, nutritious meal after working out will optimize your fat-burning potential while helping to create lean muscle tissue—a combination that can raise metabolism and keep it high.

Try not to allow yourself to get dehydrated while exercising. Most of us don't tend to drink enough while we're out jogging or using a treadmill. Dehydration is a metabolism killer, and drinking a big glass of water after exercise can help prevent or mitigate that problem. It can also curb your appetite so you feel satisfied eating smaller amounts (▶38).

6 Exercise Safely by Including Warm-Up and Cool-Down Periods

Efficient metabolism relies on building appropriate transitions into your workouts. The American College of Sports Medicine recommends a warm-up period before beginning your routine, to prepare your muscles for the stress of a workout. Warming up can also enhance metabolism. A study of male athletes, published in the *Journal of Sports Sciences* in 2001, found that as a result of warming up before intense exercise "the mechanism by which muscle temperature is elevated influences certain metabolic responses during subsequent high-intensity exercise."

You'll never see a professional marathoner roll out of bed, lace up his shoes, and start running without warming up first. That's because muscles, joints, ligaments, and other tissues can better handle the demands of a strenuous workout after they're warmed up and limber. This warm-up period helps prevent muscle soreness, as discussed in a 2007 article in the *Australian Journal of Physiotherapy*, and reduces the risk of injury, as described in a 2005 article in the *British Medical Journal*.

Fit Five-Minute Transitions into Your Workouts

Warming up usually consists of performing aerobic activity at a gentle pace. If you're planning to go for a run, for example, warm up by walking, varying your stride to test your range of motion, then move into a light jog for about five minutes. The goal is to get your heart pumping faster and to raise your body temperature, but the increase should be gradual. Doing it slowly delivers more oxygen-rich blood to your muscles at a moderate rate, which helps prepare them for the tasks ahead. The warming up period should take place immediately prior to stretching, according to the American College of Sports Medicine.

It's equally important to end your workout with a cool-down period to keep muscles healthy and prepared for their next workout. Cooling down occurs either as you're finishing your workout, or right after you've completed your cardio exercise. The goal of a cool-down period is to lower your heart rate and body temperature after exercise, but in a gradual manner. After going full-speed on a stationary cycle, or jogging at a good clip for thirty minutes, aim for a period of five to ten minutes during which you slowly reduce the intensity and pace of your workout. Jog slower until you're walking, for example.

Maintain Muscle Health

During strenuous exercise, your muscles use their glycogen stores to provide you with energy, and the body produces a byproduct called lactate (or lactic acid), which builds up over the course of an intense workout. With a proper warm-up period, lactic acid builds up gradually and then is dispersed during a cool-down period after you've finished exercising. Without a transitory cool-down period, lactic acid levels can fall abruptly, which may cause sore muscles.

According to a 2006 article in the journal *Applied Physiology*, studies show that suddenly ceasing intense exercise also leaves your heightened blood flow with nowhere to go. Blood can then pool up in arm and leg muscles, which may lead to light-headedness. A cool-down period allows energy byproducts to dissipate from the muscles slowly, and enables the body to redistribute blood flow to appropriate areas.

Proper warm-up and cool-down periods help prevent muscle strain and keep your muscles in good condition for the next day's workout. Daily exercise (combined with appropriate diet) is the best way to raise metabolism and keep it high. Without warming up and cooling down, you may find that you need a few days' rest in between each workout session.

66

Change Your Eating Schedule to Keep Metabolism High

Changing your usual eating pattern can increase your metabolic rate. To maximize metabolism, try eating a healthy meal or snack every couple of hours, instead of the customary "three squares." A 2000 study of female college athletes, published in the journal *Medicine & Science in Sports & Exercise*, showed that the women who didn't eat anything for extended periods during the course of the day had higher percentages of body fat than their counterparts who ate more regularly.

A Swedish study reported in the journal *Obesity* in 2008 showed that eating regular meals may help to prevent metabolic syndrome, a condition that often appears in obese, sedentary people and can lead to cardiovascular problems and/or diabetes.

Frequent Eating Fires Up Your Metabolism
Every time you eat, food is broken down into molecules, then delivered to various parts of the body as needed. Carbohydrates enter the bloodstream quickly and provide energy; proteins strengthen and maintain cell structures; fats are broken down and either burned for energy or stored for future use. Your metabolism has to step up its pace to accomplish these tasks; hence, eating increases your metabolic rate. By spreading out your food consumption throughout the day, your metabolism has many chances to fire itself up, and your energy levels remain more consistent throughout the day.

Eating only three times a day (or two, if you're prone to skipping breakfast) can put your metabolism into what is

colloquially called "starvation mode." When you get into a pattern of eating infrequently, your body starts conserving calories. You require a certain number of calories per day simply to keep your body functions operating. If those calories are not supplied through food, your body burns through its glycogen stores first. Once the glycogen is exhausted, the body starts to convert your lean muscle tissue into the energy it needs to keep your heart, lungs, and other organs doing what they're supposed to do.

At this point, two things are actively working against your metabolism. With fewer calories coming in, your body slows down to get the most out of those few that are available. Additionally, muscle mass diminishes—and muscle is metabolically active tissue that contributes significantly to a faster metabolic rate.

Almost any mechanical system operates more efficiently if you keep it running at a constant level, rather than repeatedly stopping and starting it. Your body—and the chemical reactions that make up your metabolism—operates best that way, too. Rather than giving your system infrequent fueling in the form of two or three large meals daily, which causes your metabolism to start and stop often, it is better to ingest small amounts of food more frequently, thereby keeping your system operating more consistently.

Sustained Nutrition Prevents Binge Eating

Spacing meals out at regular intervals throughout the day can also diminish binge eating. A 1999 study published in the journal *Appetite* showed that healthy men were less hungry if they spaced their meals out evenly throughout the day instead of eating fewer, larger meals. If you wait for hours until you reach the designated mealtime, you'll feel quite hungry by the time you finally start eating. Because it can take up to twenty minutes from the time you're full until your brain receives the "sated" signal, you may consume more food than your body needs before your brain convinces you that it's time to stop eating.

If preparing five or six small meals seems like too much effort or doesn't fit into your schedule, try eating three moderate meals plus two healthy snacks. Snacks are not the enemy. Provided you make healthy choices, snacks every few hours interspersed with small, wholesome meals can be a great way to provide your body with the regular nutrition it needs to keep your metabolism working at a constant rate (▶34).

67

Allow Ample Time for Dinner Digestion

If you're trying to raise your metabolism, eating a heavy dinner late in the evening isn't the best way to do it. The evening meal, for many people, is the biggest and most calorie-laden of the day. However, metabolic rate slows at night because your activity level diminishes while you sleep. Ideally, you want to fuel up when your metabolism is running faster (▶63, 64), as your food will be digested and utilized more efficiently then. Eating right before bed means your slower metabolism will tend to digest your food more slowly—and that may mean that a larger percentage of calories could get added to your fat stores.

However, a small study of rhesus monkeys at the Oregon National Primate Research Center at Oregon Health & Science University (published in the 2006 journal *Obesity Research*) showed that eating late doesn't necessarily make you more prone to weight gain than if you eat during the daytime. Calories are calories, regardless of when you ingest them. The real problem with eating late, according to Betty Kovacs, Director of Nutrition for the New York Obesity Research Center Weight Loss Program, is that many people find it more difficult to control their food intake in the evening, and end up consuming too many calories between dinner and bedtime.

Digest Food Completely Before Bedtime

Light metabolism-boosting evening meals are the best approach (▶35). Try to stick to lean proteins and vegetables, keeping fats and simple carbohydrates to a minimum. If you eat your evening meal at least two hours before bedtime, you'll give your body enough time to digest what you've eaten before you go to sleep. Because digestion takes energy and stimulates your metabolism, eating a light supper can make it easier to sleep, too. And as researchers at Stanford University and the University of Chicago found in separate studies in 2004, sleep deprivation may be linked to obesity (▶26, 84).

If you've already consumed a heavy evening meal, consider going for a walk after dinner to aid digestion and burn some calories. Light exercise can also help you sleep better. Intense exercise right after eating a large meal isn't recommended, however, because the blood supply that would normally go to work on digestion is being diverted to other muscles.

163

68

Linger Over Food and Lose Weight

The more slowly you eat, the less you're likely to eat. A 2008 study published in the *Journal of the American Dietetic Association* confirmed that eating slowly can lead to decreased food consumption. This is because it takes a while for your brain to tell you that you are already full—during which time you often continue to eat more than you need.

It takes about twenty minutes from the time you finish eating the number of calories required to satisfy your BMR until you start to feel full. Why such a long delay? A 1980 study published in the journal *Science* found that the signal that says you are pleasantly full actually comes from your intestines—not your stomach—so the food must enter, fill, and pass through the stomach first. Your intestines then relay a signal to the brain, which activates a satiety response in the hypothalamus, triggering a feeling of fullness that makes you stop eating.

Savor Your Meal—You'll Eat Less

A twenty-minute delay can be significant if you are eating rapidly and not paying attention to your body's signals—you could quickly down an entire day's worth of calories when a small snack might have sufficed.

A 2006 study of college women conducted at the University of Rhode Island measured eating speed and calorie consumption. Lead researcher Kathleen Melanson, director of the university's Energy Metabolism Laboratory, reported that the students "ate sixty-seven calories more in nine minutes than they did in twenty-nine minutes." The women also said they felt less satisfied an hour later when they ate quickly than if they slowed down and savored their food.

How can you avoid falling into the trap of gobbling more calories than you need? A good guideline is to take a small portion of food and eat it over the course of at least twenty to thirty minutes. Take the time to truly taste and enjoy your food. Savor the aroma, colors, and flavors. Chew thoughtfully and carefully (thorough chewing aids digestion, in addition to giving you more satisfaction from your food). Put down your fork in between bites, and use a small fork or spoon rather than a large one. Remind yourself that eating is more than a remedial process of fueling your body—it can, and should, be a multi-sensory, pleasurable experience.

When you've finished eating a small amount of food, evaluate whether or not you're still feeling hungry before you go for a second helping. Try drinking a glass of water (▶38). Wait a bit, and then reevaluate the situation. If you genuinely still feeling hungry, serve yourself another small portion of healthy foods, and, again, eat slowly.

Make Meals That Can't Be Gulped

If you think your food isn't exciting enough to warrant a full twenty minutes of chewing, dress it up. Experiment with cooking a variety of healthful foods that are colorful, varied in texture, and rich in flavor. Use red or yellow peppers instead of green, for example, or turn your entire meal into a skewered kabob. Combining a variety of textures can increase the amount of time it takes you to consume your meal—it's very difficult to gulp down a salad, for example, but a milkshake can be slurped in minutes. Set expectations for eating slowly. You won't be the first person at the table to finish, and you don't necessarily have to clear your plate. Changing your mealtime preconceptions—and your old eating patterns—can help you adopt the new, healthy habit of eating slowly.

Eating slowly in social situations can actually be easier than remembering to eat slowly when you're home alone. At a get-together or an evening at a restaurant with friends, avoid mindless eating by focusing on enjoying conversations with your companions.

69

Get Metabolism-Burning Benefits from Your Everyday Tasks

View your everyday tasks as opportunities to rev up your metabolism. You don't have to spend hours at the gym or hire a personal trainer to accomplish your exercise goals. Our bodies were designed to perform strenuous, daily, physical work. All activities that require some physical movement will burn calories and fuel your metabolism, especially when done intentionally, consistently, and vigorously. When you think of it this way, "working out" takes on a whole new meaning.

See Daily Chores as Workout Opportunities

Our agrarian ancestors used up lots of calories carrying out their daily chores—plowing fields, building fences, hauling water to wash clothes, chopping wood. For them, obesity and sluggish metabolisms never entered into the picture. For instance, thirty minutes of feeding cattle burns about 150 calories, milking cows takes 100 calories, and using a pitchfork to throw hay burns 280 calories.

You don't have to live on a farm to put this sort of pragmatic exercise program into practice, though. Shoveling snow in the winter uses 220 calories in thirty minutes, and sweeping out the garage or garden shed burns around 135 calories per thirty minutes.

Cleaning house can burn major calories. Thirty minutes of aggressive vacuuming requires about 120 calories, as does mopping or scrubbing stains off the floor. Dusting or making beds burns approximately eighty calories per half hour, and folding laundry uses up about seventy calories. Even spending twenty minutes or so putting away groceries will burn around sixty calories—and if you squat rather than bending over you'll get the extra benefit of toning your hips, thighs, and rear end.

Something as simple as working on your car (a do-it-yourself oil change, for example) will use up about 200 calories an hour. Helping a friend move furniture can burn 400 calories an hour—more if you have to climb stairs. Mowing the lawn is an excellent source of stealth exercise (▶15); using a traditional push mower requires 180 calories per half hour. Even spending a half hour preparing a meal can burn eighty calories (assuming you don't do too much "sampling" while cooking).

Everyday Activities Burn More Calories than You Think

If you have a choice between lounging on the couch and doing just about anything else, pick "anything else" to fire up your metabolism. Did you know that a half-hour of playing the trumpet burns eighty calories? Dancing around the room with your kids uses up at least 150 calories per half hour (more if you're really moving). Clearly, all forms of physical activity are good for your metabolism. Don't miss out on the ordinary, readily available ones that you can integrate into your everyday life.

Note that your body will burn fat most efficiently with sustained, high-energy activity. However, your mundane activities throughout the day can increase the number of calories you burn as well as the weight you lose. Remember that one pound (0.45 kg) of body weight represents 3,500 calories. If you can consistently burn an extra 500 calories per day for a year, that adds up to 182,500 calories—the equivalent of about fifty pounds (23 kg)!

70 Join a Support Group to Reinforce Your Goals

Joining a group of people whose fitness goals align with yours can provide the support and information you need to achieve your goals. Research presented at the International Federation for the Surgery of Obesity annual meeting held in Sydney, Australia in 2006 showed that obese people who belong to support groups and attend meetings regularly lose more weight than people who don't. Knowing you're not alone on your journey to better health is reassuring and affirming.

The Pluses and Minuses of Support Groups

Support groups consist of people who've gotten together for a common purpose. Generally, such groups allow members to share their experiences and receive encouragement from each other. Some groups consist entirely of your peers, while others may be run by a professional nutritionist, physician, or fitness expert. Members also exchange resources, remedies, and advice regarding their common objective. Meetings can be held anywhere—at a medical facility, religious center, YMCA, or someone's home.

The upside of being in a support group for metabolism boosting and weight loss is that you'll have an opportunity to meet people who are tackling the same issues you are. Someone may share excellent advice about raising BMR, for example; someone else may provide recipes for losing weight that really work. You're virtually guaranteed of meeting people who understand what you're going through, and you'll be able to discuss the personal details of your diet and exercise plan (things you might not even feel comfortable telling your spouse or friends).

What's the downside to joining a support group? "Corporate" groups that also include personal consultations with experts can be expensive, and the rules for participating are usually stringent. Local groups may be more flexible, and smaller meetings may be less intimidating if you're looking for a low-key setting.

In-person support groups are not for everyone. Some people just aren't comfortable sharing private matters with others, and the thought of having to participate in a group session is enough to make them cringe. If you're already sensitive about your body, sitting around talking about it with a bunch of strangers may be the last thing you want to do. If that sounds like you, consider an online support group that doesn't require face-to-face meetings.

Finding a Group that's Right for You

When seeking a support group for weight loss, fitness, raising metabolism, or anything similar, start by checking with your doctor or health-insurance office. Some medical plans may cover costs associated with belonging to a health-related support group. Most gyms and health clubs offer support groups, as do many corporate offices, so talk to the fitness director of your club or the human resources director at your place of business. Your area hospital may be able to direct you to a group in your vicinity. You can also try calling your local parks and recreation department for listings of free community support groups, or visit the recreation center directly and check the bulletin board postings. Local colleges and universities might offer similar opportunities. Your library could be a resource, too.

One of the best ways to stick with your fitness plan is to team up with a friend or family member who's working on getting healthier, too. Make a pact to check in with one another daily, by phone, email, or in person, to support and encourage each other. If you'd prefer to join an online support system, you'll find many good ones available. You can log onto a website or blog, look for a bulletin board, or join a mailing list to locate individuals with similar goals.

No, a support group won't exercise for you, cook your meals, or design a metabolism-raising plan for you. You'll still have to do the brunt of the work yourself—and the resulting sense of achievement will be yours to cherish. However, a support group can be a beneficial part of the lifestyle and habit changes you're choosing to make. And it can provide a comfortable place for you to share your successes and overcome your setbacks.

PART VI

Metabolism Killers:
What to Avoid, and Why

Don't Depend on Diet Pills

It can be tempting to take a pill that promises to let you lose thirty pounds (14 kg) in thirty days without exercising. Quick fixes, however, rarely live up to their billing. If it sounds too good to be true, it probably is. Although many people who try diet pills lose some weight initially, they don't usually lose a lot, and what they do lose is easily regained when they stop taking the pills, as shown by a 2007 review article published in the *British Medical Journal*. Some diet pills can even be dangerous.

The Dangers of Blocking Natural Signals

There are several categories of weight-loss supplements, and they work in different ways. The most common weight-loss pills reduce appetite by interfering with normal hunger signals sent by the hypothalamus (the part of the brain that connects the nervous and endocrine systems). The hypothalamus is also responsible for regulating body temperature, sleep, fatigue, and thirst. Appetite suppressants, therefore, can cause a range of unpleasant side effects such as exhaustion, dry mouth, irritability, and stomach upset as well.

Some diet pills called "fat blockers" work by blocking fat absorption in the intestines. This process can be uncomfortable, and many users report gas, loose stools, and intestinal cramps. Other weight-loss supplements work by stimulating the nervous system. Although their goal is to increase your rate of digestion and metabolism, they can also raise blood pressure; this carries an increased chance of heart attack and cardiac disease. Diet pills containing ephedra were banned by the Food and Drug Administration in 2004 because they increased the risk of heart attack and stroke, as reported in 2003 in the *Journal of the American Medical Association*. Another controversy erupted in 1997, when it was discovered that there was a correlation between patients who took the appetite suppressants fenfluramine and phentermine and heart-valve disease, as reported in 1997 in the *New England Journal of Medicine*.

Many weight-loss supplements can be disastrous for people trying to raise their metabolisms. Anything that suppresses appetite can interfere with your body's natural response to hunger and thirst signals. Feeding the body a constant stream of nutrient-dense foods is one of the best ways to keep metabolism high

(▶**34, 66**). Lowering caloric intake due to an artificially suppressed appetite can actually slow metabolism, which leads to slower weight loss and decreased energy. Sufficient sleep is also extremely important for optimal metabolism (▶**26, 67**), and some diet pills can disrupt normal sleep patterns.

Excitement Gives Way to Disappointment

Consider the emotional risk as well. After investing in a new diet pill, you may feel excited and hopeful that this new "cure" will finally enable you to shed that unwanted weight and turn you into a fat-burning machine. But when a particular pill's promises are not realized, disappointment and frustration may ensue. Negativity (▶**73**) can lead to emotional eating or cause you to give up altogether. In fact, it appears that the people who experience success with diet pills are the same ones who are sincere about dieting and exercising, instead of relying totally on supplements to take the weight off.

No pill is a substitute for healthy eating habits and exercise, and none will elevate your metabolic rate or reduce weight and body fat long term, in the absence of exercise and proper diet. The only proven way to raise metabolic rate is through permanent behavioral changes. Some supplements can certainly help raise your metabolism and shed unwanted fat—but don't confuse necessary vitamins, minerals, and herbs with "cure-all" potions that may do more harm than good.

173

72

Don't Be Fooled by Artificial Sweeteners

If you think that drinking diet soda and cooking with artificial sweeteners is a short cut to weight loss and metabolism management, think again. Artificial sweeteners may save you a few calories, but in the long run they can lead to weight gain, insulin resistance, and a sluggish metabolism. A 2008 animal study published in the journal *Behavioral Neuroscience* found that consuming artificially sweetened foods increased appetite, caloric intake, and weight gain.

There are several popular artificial sweeteners, or sugar substitutes. These additives use something other than white table sugar (sucrose) to make foods and drinks taste sweet. Aspartame (marketed as NutraSweet and Equal) is scientifically known as aspartyl-phenylalanine 1-methyl ester—quite a mouthful! Saccharin (found in products like Sweet 'n' Low) is benzoic sulfinide. Sucralose (marketed as Splenda), neotame (produced by NutraSweet), and acesulfame potassium (marketed as Sweet One and Sunett), are other readily available, non-sugar sweeteners.

Confusing Messages Create Long-Term Problems

Why are sugar substitutes bad news for metabolism-boosters? When you eat something sweetened with sugar, your brain gets an early heads-up and prepares your metabolism to start digesting glucose. Artificial sweeteners also induce the brain to send "here comes something sweet" signals. But because sugar substitutes don't follow up with the calories of glucose, your metabolism can falter as it tries to sort out those signals. The hypothalamus, the part of the brain that gauges appetite, may eventually become less effective at accurately reporting hunger and satiety.

Both sugary and artificially sweetened foods can lead to a variety of dangerous conditions. Insulin resistance occurs when the body isn't as sensitive to insulin as it should be. Fat, muscle, and other cells become resistant to insulin, so more insulin is required in order for the body to function properly. Studies, as reported in a 2007 paper published in the *Journal of Nutrition*, show that consuming sugary drinks can lead to insulin resistance, which often precedes diabetes. A 2008 paper published in the journal *Circulation* further showed that people who regularly drink diet sodas have an increased risk of obesity and "metabolic syndrome" (the name for a group of conditions, including insulin resistance and high blood pressure), both of which can be a precursor to diabetes or heart disease.

Hit the Sweet Spot with Natural Flavors

So what should you do when your sweet tooth is crying for satisfaction? Rather than loading up with artificially sweetened drinks and foods, ingest a small amount of the real thing. Instead of a diet soda, try sparkling water with a splash of fruit juice. Ditch the "diet chocolate" bar and eat a square of rich, real dark chocolate, which contains a variety of healthful nutrients.

Alternatively, try a non-sugar natural sweetener like Stevia. Derived from a South American plant, it contains no calories and doesn't raise blood sugar. It's significantly sweeter than sugar, so you'll only need about one-tenth as much in recipes that call for sugar. Honey and succinate are other sweet alternatives, but they both contain calories and affect blood sugar.

Generally, the best guide for regulating metabolism is "go natural." Splurge occasionally with a small glass of pure fruit juice, such as fresh-squeezed orange juice. Or try an exotic new fruit—how about a durian, a matisia, or a soursop? Shake up your routine (▶29) and quench your cravings with something new, healthy, and free of artificial ingredients. Once you get used to the natural taste of foods, you may find your cravings for diet soda and artificially sweetened treats have dissipated.

Artificial sweeteners may have their place in certain people's diets. If you are diabetic and must closely monitor your sugar intake, careful use of artificial sweeteners may allow you to enjoy baked goods and other goodies occasionally. As with all food choices, remember that moderation is everything.

73

Maintain the Right Attitude to Sustain Good Habits

Fitness isn't just physical—it's mental as well. Don't forget a vital aspect of raising metabolism: your attitude. A poor or destructive attitude about yourself can slow progress measurably. According to Pamela Peeke, M.D., author of *Fit to Live*, you need to "cut the mental fat, and that will lead to cutting the waistline fat." A study published in 2001 in the *Journal of the American Dietetic Association* found that mothers who had positive "can-do" attitudes about losing weight were more likely to exercise after giving birth and lost on average twelve pounds (5.5 kg) more than those who didn't.

Making an Emotional Commitment, for the Long Term

For most people "going on a diet" implies that a set of rules will be adopted specifically to lose a certain amount of weight. Studies, such as a 2007 review published in *American Psychologist*, confirm that most people do not lose permanent weight from going on a diet. Changes that will raise metabolism, such as eating and exercising correctly, are lifestyle changes. They are not temporary additions or subtractions from your existing way of life, and can't be thought of as "something I'll do this week" or "just for today." The adjustments you'll make will be permanent, if you want to raise metabolism and keep it high for life. They'll become a part of how you see yourself and the world around you.

Having a positive attitude about metabolism can also lead you to actively seek out healthy ways to raise it. A 2007 article published in the *Journal of Experimental Social Psychology* explains that people tend to look for information that matches their attitudes. If you know that you're doing the right things and feel good about the progress you've already made, you may be more likely to seek out new information that can benefit you further.

An emotional commitment to raising metabolism is a prerequisite to actually losing the weight and building the muscle. If you're not ready to commit to lifestyle changes, or if you think it's just too inconvenient to alter your eating and exercise habits for more than a short period of time, you're putting up a barrier to your success. Commit (verbally, mentally, and/or in writing) to the notion that the changes you make will determine your outcome (▶**59, 61, 62**).

Building on Your Success—and Setbacks

"Look at the patterns and habits in your life that you are dragging around with you that get in the way of success," Dr. Peeke points out. Rather than blaming yourself for eating too much last night, stick to a healthy meal plan today. Instead of lamenting the fact that you didn't work out yesterday, go to the gym today. Focusing on your mistakes is self-defeating. Everyone has setbacks; the trick is to use them as guides for making changes that will lead to success.

Additionally, make a point of choosing your words carefully, even the words you don't say out loud. Everything you say and think should support your underlying goals of raising metabolism and adopting healthy lifestyle changes. Thoughts like "Boy, do I feel fat today" or "I'll never be able to lose all this weight" undermine your willpower and your self-image. Verbalizing your perceived failure makes it seem real, permanent, and insurmountable.

You may wish to speak with a professional about your attitude toward weight loss, metabolism, and health in general. If you don't feel comfortable talking to your regular physician, locate a qualified mental-health professional, particularly one who specializes in issues of fitness and body image.

4 Steer Clear of Saturated and Trans Fats

The F word in the lexicon of diet and metabolism is Fat. Fats are the foods we're most often told to eliminate from our diet because eating fat makes us fat—right? Actually, trans fats and saturated fats are the ones to steer clear of when you're trying to raise metabolism or lose weight. Studies, such as one presented at the 2006 Scientific Sessions of the American Diabetes Association, link the consumption of trans fats with weight gain, even in diets with restricted caloric intake.

Trans fats (also called trans fatty acids or partially hydrogenated fats) are a type of unsaturated fat created through the partial hydrogenation of vegetable oil. By adding hydrogen to unsaturated fat, the fat becomes saturated, meaning it is more solid and stable for cooking and baking at higher temperatures. Trans fats are commonly found in shortening, margarine, fried foods, cookies, chips, doughnuts, pies, and other pastries and snack foods. They are used frequently because they're inexpensive, have a long shelf life, and can be used more than once (such as in the fryers found in some fast-food restaurants). As of 2006, the FDA has required all food manufacturers to list on their packaging the amount of trans fat a product contains, so it's relatively easy to avoid.

Finding the Right Substitute for Saturated Fat

Saturated fats, or saturated fatty acids, are a kind of fat that remains solid at room temperature. Animal and a few plant products are the main sources of saturated fat. Specific sources include coconut oil, butter, lard, full-fat dairy products, and fatty cuts of meat. Many commercially available fried or processed foods contain plenty of saturated fat. Saturated fats are very high in calories.

Much contradictory evidence exists concerning the relationship between consuming saturated fats and weight gain, as well as the connection between eating fat and high cholesterol. The Centers for Disease Control and Prevention report that in recent years, American men have decreased their intake of saturated fat by 14 percent.

At the same time, obesity is on the rise. Why? The answer may well be that this decline has been accompanied by an increase in carbohydrate consumption. Although cutting back on saturated

fats to reduce overall caloric intake may be beneficial, if you replace those fats with carbohydrates you're likely to be in worse shape, explains Richard Feinman, director of the Nutrition & Metabolism Society and Professor of Biochemistry at the State University of New York Downstate Medical Center in Brooklyn.

Metabolism-Boosting Alternatives

Not all fats are created equal. Mono-unsaturated fats and polyunsaturated fats, such as omega-3 fatty acids (▶ 21), are a great energy source and are essential to human growth and development. These healthy fats can contribute to a weight-loss and metabolism-boosting diet.

When in doubt, usually less is more, and natural is better than processed. A small amount of saturated fat or trans fat may be okay, but whenever possible, choose unprocessed products like butter over those that contain processed trans fat, such as margarine.

75 Just Say No to Refined Sugars

Sugar may provide a burst of quick energy, but it also gives you lots of empty calories. According to USDA recommendations, you shouldn't consume more than ten to twelve teaspoons of sugar—any kind of sugar—per day. However, in 2000, Americans averaged about three times that amount. Many professionals in the medical and nutrition fields oppose ingesting any refined sugar, beginning with Dr. William Coda Martin more than sixty years ago. In an article published in 1957 in the *Michigan Organic News,* Dr. Martin called sugar a poison, devoid of all nutrients.

A Lot Removed and Little Gained
Although it's true that foods containing simple sugars (such as simple carbohydrates) are quickly converted into energy, these foods provide more sugar than you can use at the time of ingestion. Carbohydrates that aren't immediately utilized for energy can be converted to glycogen (the sugar stored within muscles and the liver) for future use. But storage space is limited; any extra sugar is stashed as body fat. Remember, a pound (0.45 kg) of lean muscle raises metabolism faster and burns three times as many calories as a pound (0.45 kg) of fat.

Refined sugars are created when sucrose is extracted from source plants, namely sugar cane and sugar beets. In the refining process, most of the nutrients from the raw plants are removed; white table sugar is usually bleached and further filtered. White sugar, brown sugar, cane sugar, and powdered sugar are all types of sucrose.

Dextrose (also known as corn sugar) is another type of refined sugar, created from cornstarch. High-fructose corn syrup is another cornstarch derivative. According to an article published in the *San Francisco Chronicle* in 2004, "Almost all nutritionists finger high-fructose corn syrup consumption as a major culprit in the nation's obesity crisis."

High-GI Sugars and Insulin Production
One of the most serious implications of refined sugars for metabolism and weight control involves insulin, the hormone that helps tissues to absorb and use glucose. Insulin also plays a vital role in regulating metabolism and digestion. Foods high in refined sugars rank high on the glycemic index, which means that they quickly raise blood glucose levels. Conversely, foods with

a low glycemic index (milk products, beans/legumes, and many fruits and vegetables) are digested more slowly, so your body has more time to absorb and process the sugars (▶16). A 2004 study published in the *Journal of the American Medical Association* showed that reducing your glycemic load, by consistently eating foods that are low on the glycemic index, can help regulate metabolism and promote weight loss.

The pancreas increases insulin production whenever too much sugar is ingested; this insulin helps supply muscle and liver cells with glycogen. When those cells are full, the extra glucose stays in the bloodstream, which causes the pancreas to make more insulin; repeated over and over, this process leads to increased fat storage (read: sluggish metabolism). It can also cause the body to develop insensitivity to insulin, eventually producing the condition called "insulin resistance."

A Hard Habit to Break

Eliminating the refined sugars from your diet can be difficult, as research shows that sugar can be more addictive than drugs. A 2007 article published in the journal *PLoS One* showed that when rats were given a choice between cocaine and sugar, 94 percent of the rats chose sugar. Reducing the quantity of refined sugars you eat, however, can help prevent weight gain and a host of other health complications. Many processed foods contain a refined sugar somewhere in the ingredient list, so always read labels. Use natural sweeteners where possible. Raw honey, for example, is digested more slowly than refined sugars, has enzymes that aid digestion, and contains far more nutrients than refined sugar (however, it is still high in calories).

Sugar substitutes aren't the answer, either (▶72). Try unsweetened breakfast cereals rather than sweetened ones, and put sliced strawberries or fresh blueberries on top instead of sugar. Nix soft drinks (▶78). Choose 100 percent fruit juice rather than juice-flavored drinks, which tend to contain large amounts of high-fructose corn syrup. Fruit juice still has lots of natural sugar, though, so if you're counting calories drink it in small quantities and quench your thirst with water.

76

Avoid Refined Flour to Increase Power

Most of us don't have time to bake our own bread, much less grind wheat into flour. However, bread and baked goods made with refined flour are far less metabolism-friendly than those that contain whole grains (▶16).

Consumption of products made with refined flour has been linked to weight gain and increased body fat, suggests a 2007 study published in the journal *Obesity*. Increased fat supplies can result in higher levels of body fat, which leads in turn to a slower metabolism. In a 2006 study published in the *American Journal of Clinical Nutrition*, participants who had diets low in whole grains showed much higher incidences of fatty liver disease, a recognized complication of obesity marked by fat deposits in the liver.

More Gain with Whole Grain

The wheat-refining process eliminates the bran and germ components of the wheat kernel, and most of the wheat's important nutrients as well. "Whole-meal flour" or "whole-grain wheat flour" uses the entire wheat kernel. (Something simply labeled "wheat flour" doesn't necessarily contain the whole grain—what we call "white flour" is made from wheat.) In 2003, a Minnesota study of adolescents drew a connection between lower BMIs (see p. 14) and eating whole grains. In the same year, Harvard scientists found that women who ate whole grains had a lower risk of weight gain.

Many countries require nutrients to be added back into flour (which explains why many refined flours are labeled "enriched" or "fortified"). These nutrients don't replace what was in the original wheat kernel in the same proportions, though, nor are they digested and absorbed as well as the real thing.

Like refined sugars, refined flours were designed to make our lives easier. However, both can have devastating effects on weight, metabolism, and general health. Refined flour is broken down in the digestive tract into simple sugars, which are quickly absorbed into your system and converted to glycogen to be used for energy. Any excess gets stored as fat. Refined flour and sugar together—found in most breads, cereals, pastries, snacks, and other commercially prepared products—perform a double whammy on your metabolism.

Flours with High Nutritional Value

What are some metabolically advantageous alternatives to refined wheat flour? Complex carbohydrates are always a better bet than simple ones (▶16). Consider rye flour, which is created from rye grass and used in rye and pumpernickel breads. Cornmeal is another choice that, generally speaking, contains more nutritional value than refined wheat flour. The nature of the corn-refining process leaves more of the nutrients intact. Buckwheat flour also fares well during the refining process. Almond flour, coconut flour, chickpea flour, and rice flour are some other good substitutes, though they all have different consistencies and tastes from traditional wheat flour. You can also add wheat germ, an extract made from the nutrient-rich part of the wheat kernel, to cereal or yogurt, or mix it in with flour when you make your own baked goods.

Steer Clear of Whole-Milk Dairy Products

Eating dairy products is one of the best ways to get calcium into your diet. But did you know that low-fat (1 to 2 percent fat) milk actually provides more calcium and protein than whole milk? People who eat diets high in calcium tend to weigh less—and lose weight more easily—than people with a lower calcium intake (▶23). A 2001 article published in the *Journal of the Federation of American Societies for Experimental Biology* showed that calcium also has a protective effect against the loss of lean muscle mass.

Fat and Calorie Content in Milk Varieties

The fat content of different kinds of milk varies considerably, as shown in the table at right, but whole milk and products made from it—cheese, yogurt, and ice cream—are especially rich with fat and calories. Raw (unpasteurized) milk has an even higher fat content of at least 4 percent.

Calcium Leads the Way in Burning Off Fat

A 2000 study presented at the Experimental Biology meeting in San Diego showed that low-fat, high-calcium dairy products had a significant effect on fat burning in laboratory mice. In the study, overweight mice fed a diet that included low-fat dry milk lost 69 percent of their body fat, as compared to mice fed calcium supplements, which had a decrease of 42 percent in body fat. Mice whose diets were not supplemented with calcium only lost 8 percent of their body fat.

The study suggests that low-fat dairy products—and the calcium in them—help adjust fat burning in cells to allow you to burn fat and lose weight without the metabolic suppression that often comes from caloric restriction.

Type of Milk	Fat (%)	Calories per Cup	Fat per Cup (grams)
Whole milk	3 to 3.5	150	8
Reduced fat	1.5 to 2	120	4.5
Low-fat or skim	1	100	2.5
Non-fat	0	85	0

Unless you're committed to a vegan diet or lactose intolerant, eating dairy products can help keep your metabolism high by providing calcium, magnesium, and protein (▶17, 23). Note that non-fat milk has the same amount of calcium as whole milk, so if you are trying to lose weight choose low- and non-fat dairy products. Switching from whole to lower-fat milk can save fifty to seventy calories and up to eight grams of fat per serving. If the difference in taste is a deal-breaker, try mixing half whole and half skim products for a few days before switching entirely to low-fat varieties.

Dodging the Dairy Diet-Busters

Milk probably isn't the only dairy product in your diet. Cheese makes up a sizable component of diet—about thirty pounds (14 kg) per person annually in the United States, according to the U.S. Department of Agriculture (nearly three times the amount consumed in 1970). Cheese tends to be higher in fat than milk. Cheddar is about 50 percent fat, and contains about ten grams of total fat per ounce (28 g); feta and mozzarella are slightly better, with about six grams of total fat per ounce (28 g), whereas whole-milk ricotta is one of the biggest diet-busters, with sixteen grams of fat per ounce (28 g). Some cheeses come in low- and non-fat varieties, so choose carefully and control portion size.

Butter typically is about 80 percent fat—one tablespoon of butter contains 100 calories and eleven grams of fat. Stick margarine has similar fat and calorie values, whereas tub margarine contains fewer calories (sixty per tablespoon) and less fat (six grams). The best bet? Use butter and butter substitutes—which are often high in trans fats (▶74)—sparingly, if at all. Also consider a lower-calorie, low-fat alternative such as cream cheese—even the whole-fat version only has about half the calories and fat of butter.

Take a close look at your dairy consumption, as high-fat dairy products may be tucked away in the corners of your diet. That dollop of cream in your morning coffee, for example, can have a fat content of 30 to 40 percent and contain five grams of fat and fifty calories per tablespoon. "Half and half" (half milk, half cream) has a lower fat content, around 10 to 15 percent, but still contains about 20 calories per tablespoon. By comparison, a plain non-dairy creamer only has half a gram of fat and ten calories per tablespoon.

Eliminate Soft Drinks from Your Diet

Drinking your calories may be worse for your waistline and metabolism than eating them. "It seems that when you drink your calories as opposed to eating them, your body may not sense that you've just taken in those calories and your appetite doesn't seem to compensate," Caroline M. Apovian of the Boston University School of Medicine wrote about a study published in the *Journal of American Medical Association* in 2004. "The appetite circuit might not be programmed to register liquid calories."

The study of fifty thousand nurses found that drinking just one twelve-ounce (360 ml) sugar-sweetened soda per day caused the women to gain weight. Nurses who consumed a soft drink a day had an average weight gain of 10.3 pounds (4.7 kg), whereas those who drank one (or fewer) per week put on less than three pounds (1.4 kg) over a four-year period. This result was thought to be not only due to the extra calories in the soft drinks, but also from the ingestion of large amounts of sugars that the body absorbed easily and converted directly to fat (▶75).

Liquid Calories Add Up Fast

Actually, your risk of weight gain from consuming all those empty calories could be far greater. Do the math: A twelve-ounce (360 ml) can of Pepsi contains 150 calories. If you drink one can per day, that's a total of 54,750 calories in a year. One pound (0.45 kg) of fat equals 3,500 calories. In one year, what may seem like a minimal amount of soda will add 15.64 pounds (7 kg) without contributing a single bit of nutrition.

A 2008 article published in the *Journal of Urban Health* also linked drinking soda daily to weight gain and higher BMI, as well as a heightened risk of diabetes and heart disease. A 2007 article in the journal *Circulation* showed that drinking one soda per day increased your chances of developing metabolic syndrome, a condition associated with obesity and lack of exercise, by 50 percent.

Carbonation Depletes Reserves of Calcium

Carbonated drinks were originally produced by naturally carbonated mineral springs; their artificially produced offshoots have been around since the eighteenth century. Sweetened versions have gotten more potent over the years as cane sugar has been replaced by corn syrup. Switching to artificially sweetened "diet" sodas isn't the answer, though. A 2008 animal study published in the journal *Behavioral Neuroscience* found that ingesting artificial sweeteners increased appetite, caloric intake, and weight gain (▶72).

Even the carbonation in soft drinks can impair weight loss and metabolism. Carbonation can reduce the amount of calcium stored in the body. The phosphoric acid in sodas also contributes to calcium loss. People who eat diets high in calcium tend to weigh less—and lose weight more easily—than people with a lower calcium intake. A 2005 study of people trying to lose weight, published in the journal *Obesity Research*, indicated that test subjects who consumed at least 1,200 milligrams of dairy calcium per day burned substantially more fat from around their waists than people who consumed the same number of calories, but less calcium (▶23).

Make a Change for the Better

If you're trying to ditch the soda habit, be sure you're not replacing it with something just as bad. Commercially prepared sweetened iced teas can contain just as many calories as sugary sodas. Fruit juice blends, fruit punch, and juice drinks are usually not carbonated, but they often have as much as 130 calories and thirty grams of sugar per eight-ounce (240 ml) cup. Sports drinks contain electrolytes, which can help replenish the body's stores after a long, intense workout, but few of us need the 200 calories or thirty grams of sugar per eight-ounce (240 ml) serving that many of these drinks contain. What about "flavored waters"? These have about half the calories of soda, and many replace sugar with artificial sweeteners, which can do real damage to your metabolism, too (▶72).

Completely eliminating soda is one of the best ways to cut calories from your diet. If you choose to enjoy a soft drink occasionally, consider it a treat. Think of soda like dessert. You don't want to drink it all day, every day. If you are thirsty, drink water. Water quenches your thirst and provides metabolism-boosting hydration (▶37).

79

Minimize Salt Intake to Avoid Water Weight Gain

If you're exercising regularly and reducing your caloric intake, yet you still look and feel bloated, it is possible that the villain could be salt. Too much sodium in your diet causes your body to retain water, increases fat-cell mass, and contributes to weight gain.

Research by professors Dr. Heikki Karppanen of the University of Helsinki and Dr. Eero Mervaala of the University of Kuopio, both in Finland, which was published in the journal *Progress in Cardiovascular Diseases*, suggests that increased salt consumption has contributed to escalating obesity all over the world. According to the American Salt Institute, for example, Americans upped their salt intake by more than 50 percent between the mid-1980s and the late 1990s.

The U.S. recommended daily allowance of sodium is 2,300 milligrams per day, though some estimates, as discussed in a 2005 article in the *International Journal of Epidemiology*, place the actual required need for sodium at closer to 250 to 500 milligrams per day. Five hundred milligrams is about one-quarter of a teaspoon per day—and that includes what's found in drinks, prepared foods, and snacks. It's safe to say that most people consume significantly more sodium than they need.

Sodium Intake and Fluid Retention

Salt and sodium, by the way, are not the same thing although we often use the terms interchangeably. Sodium is a mineral, "Na" on the periodic table. Table salt is sodium chloride, or NaCl; it contains about 40 percent sodium.

Although there are many other types of salt, when you consider reducing the amount of sodium in your diet you're usually thinking of everyday table salt.

After eating foods that are high in sodium, the body develops an electrolyte imbalance. Electrolytes are electrically charged ions; in this context, think of them as substances commonly found in the body (such as potassium, sodium, magnesium chloride, and calcium) that work to help transfer electrical impulses (such as muscle contractions) between cells. The natural response to this imbalance is thirst, which urges you to consume enough liquids to bring electrolyte concentrations back into balance. As more sodium-laden foods are eaten, the thirst queue increases correspondingly. The body excretes excess sodium through the kidneys, but this process is not instantaneous, and water retention can result. Avoiding excessive sodium intake is the key to preventing this type of fluid retention.

Excess sodium in the diet can affect metabolism as a secondary condition. When you take in more sodium than you can effectively metabolize, you retain fluids. For most of us, our bodies have counter mechanisms that prevent very significant fluid retention following high sodium intake. If your kidneys are not working at full capacity, your tissues will start to swell from all that excess fluid. Swollen joints tend to feel stiff and may impair your movement, so you become disinclined to exercise. Skipping exercise can lead to a loss of muscle tone, increased fat stores, and a generally slower metabolism. See where this runaway train is headed? Derail the problem by limiting your sodium intake.

Screening for Salt in Prepackaged Foods

Other than adding table salt to food during or after the cooking process, where does all the sodium in our diets come from? Salted chips, nuts, pretzels, and other snacks are obvious culprits, but they're not the only ones. Processed lunch meats are very high in sodium, as are many canned soups, cheese, rice mixes, and prepackaged meals. Ketchup,

barbecue sauce, and other condiments can also contain lots of sodium. Even drinks, particularly "energy formulas," may include large amounts of sodium— read package labels carefully to be discover how much sodium you're actually consuming.

For most people, reducing the amount of sodium you consume is a sensible step toward a healthy diet, but don't go overboard. Sodium is, after all, vital for good health. When taken to extremes, sodium deficiency can cause serious problems. For example, a serious concern for distance athletes can be hyponatremia, an electrolyte imbalance characterized by low concentrations of sodium in the body fluids that in extreme cases is life-threatening.

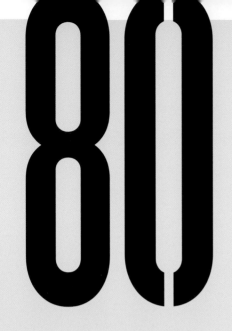

Limit Alcohol Intake to Reduce Your Waist Measurement

If you're accustomed to enjoying a "happy hour" cocktail after work or a glass or two of wine with your dinner, you may be downing lots of empty calories that can pack on weight and slow your metabolism. Pure alcohol is quite high in calories—about seven calories per gram. It follows, therefore, that your favorite alcoholic beverages are also high in calories.

A 12-ounce (360 ml) beer contains about 150 calories ("lite" beer has about 100 calories), a four-ounce (120 ml) glass of red wine has about 80 calories, and wine coolers usually have around 150 calories per six-ounce (180 ml) serving. Gin and vodka contain around 65 calories per ounce (30 ml); mixers made with fruit juice or soft drinks add even more calories. A martini will give you about 150 calories per four-ounce (120 ml) serving and a margarita can have up to 500 calories. Liqueurs generally contain between 130 and 200 calories per 1.5-ounce (45 ml) serving.

What About That Beer Belly?

Drinking alcohol may adversely impact your metabolism in other ways, too. A study reported in 1999 in the *American Journal of Clinical Nutrition* showed that men who drank two vodka cocktails made with sugar-free lemonade during a half-hour period experienced a 73 percent decrease in their fat metabolism.

And you've heard of a "beer belly," right? People who drink alcohol on a regular basis may have more trouble reducing their percentage of body fat than their counterparts who do not drink alcohol. For example, an article published in 1998 in the *International Journal of Obesity and Related Metabolic Disorders* revealed that increasing levels of alcohol consumption were associated with increasing waist-to-hip ratio measurements among the study participants, indicating greater amounts of deposited abdominal fat.

Consider the way alcohol itself is metabolized. Alcohol is not digested in the same way as other foods and drinks. Calories from alcohol are burned quickly, before carbohydrates, proteins, and fats. This means more of those nutrients are shuttled toward fat production, so fat has more of a chance to accumulate and remain stored in your body. In 2004 the *International Journal of Obesity* published a study comparing the effects of drinking 150 calories of wine a day against the effects of consuming 150 calories of grape juice. Test subjects consumed the same number of total calories per day (1,500). Over a three-month period, however, the group who drank grape juice lost, on average, 20 percent more weight than those who drank wine.

Alcohol's Other Metabolism-Busting Properties

Alcoholic drinks generally contain few nutrients, which is why they are often categorized as "empty calories." You may have heard that some experts agree moderate drinking—usually considered to be the equivalent of one four-ounce (120 ml) glass of wine every day—may have cardiovascular benefits; a 2005 article in the journal *Circulation* showed a link between moderate red wine consumption and a lower risk of heart disease. However, there are clear downsides to drinking alcohol if your goal is to speed up your metabolism. Alcohol is a diuretic (a substance that increases urine output), meaning that alcohol actually dehydrates the body. Research conducted in 2003 at the University of Utah showed that dehydration leads to a slower metabolism (▶37). That means it's a good idea to follow alcoholic beverages with plain water to ensure that your body remains adequately hydrated.

Another reason why people who drink alcohol may have a difficult time losing weight is that alcohol increases appetite and relaxes inhibitions. After a few drinks, you may end up eating foods that you would normally avoid. If you've already eaten your target number of calories for the day, a glass of water is a smarter choice than wine or beer.

Alcohol may let you fall asleep faster, but you might sleep less deeply and for a shorter period of time. And as researchers at Stanford University and the University of Chicago showed through studies conducted in 2004, sleep deprivation may be linked to obesity (▶26, 84).

81

Resist the "Quick Fix" of Diet Foods

Are you too busy to plan healthy, calorie-conscious meals? Substituting quick-and-easy "meal replacements" in the form of shakes, bars, or powders that profess to provide all the nutrients you need can be attractive to busy people. Although these formulations may simplify your life, prepackaged "diet meals" can actually promote overeating.

It's easy to consume a diet bar or weight-loss shake in a few moments, but it takes your brain up to twenty minutes to register that you're feeling full. In the meantime, you're likely to keep eating even though you've already consumed all the calories allotted for that meal. Studies, such as one published in 2003 in the *Nutrition Journal*, show that meal replacements can aid weight loss, but only if the replacements are continued

long-term, as demonstrated further by a study published in 2008 in the journal *Obesity Management*.

The Satisfaction Factor

It's easy to down these quick-fix foods when you're on the run, barely noticing that you've consumed your calories for that meal. Eating on the run often leads to overeating. Patricia Pliner, Ph.D., a professor of psychology at the University of Toronto Mississauga, led a study that compared the effects of standing or sitting down to eat a meal. One group of participants ate while standing at a kitchen counter, the other group sat at a table to dine. Researchers monitored what both groups ate at their next meal and discovered that those who ate standing up consumed 30 percent more than those who sat down to eat. As a

result of the study, Dr. Pliner advises, "Don't eat on the run. It won't feel like a meal, and you may subconsciously grant yourself permission to eat more later."

Additionally, most diet foods and bars do not contain sufficient unprocessed, whole foods to encourage satiety. Your daily recommended allowance of fiber and vitamin A may be supplied by a meal bar, but what's missing is the satisfaction you get from eating adequate portions of fresh vegetables, fruits, and grains. Without being able to savor a variety of tastes and textures, you're likely to still feel hungry—emotionally and physiologically—after eating a diet bar or drinking a meal replacement shake. A prepackaged allotment of rations may fuel the body, but it's apt to leave you wanting more.

Meal Substitutes Make False Promises

Many meal replacement bars and shakes make false promises that can undermine your metabolism-raising plans. If you see something labeled "diet bar" or "weight-loss shake," the presumption may be that eating these foods will make you lose weight. Any food that gives the impression of causing weight loss or boosting metabolism, without being accompanied by exercise and other lifestyle changes, can potentially cause disappointment and setbacks.

Does a bar or shake give you more than you're looking for? The composition of individual bars varies widely, so make sure you read ingredients and nutritional information carefully. Meal replacement bars generally contain 200 to 250 calories; some contain ten to twenty grams of protein, with other vitamins and minerals added in. On the downside, though, they may have ten grams of fat, thirty grams of carbohydrate, and substantial amounts of sodium and sugar. Some also include artificial sweeteners, colors, or flavors, and they are usually packed with preservatives and other highly processed ingredients. By comparison, a salad with vegetables, a low-calorie dressing, and a few slices of chicken breast might also add up to 200 calories, but you'll be getting the full, unprocessed nutrients of your foods—and the satisfaction of eating a proper meal.

Meal substitutes may have their place in certain people's diets, however. Nutritional shakes can be a lifesaver for people who cannot tolerate chewing or swallowing solid food, for example, and a protein bar is usually a better choice than fried fast food when there are no other options for a quick meal. In general, though, the compelling arguments in favor of meal replacements appeal to your sense of time and convenience rather than good sense.

The goal of any permanent metabolism-raising plan should be to learn how to eat correctly and healthily, not to establish a reliance on meal substitutes. Meal replacements offer a quick fix; they don't require that you learn anything about making good food choices or changing your lifestyle and exercise patterns permanently. And a large part of regulating metabolism is adopting healthy habits that you can maintain for a lifetime.

82

Be Alert to the Effects of Incidental Eating

Even though you may be preparing metabolism-friendly meals and working out regularly, unplanned calories can sneak into your diet almost unnoticed—until you step on the scale. Free food and absent-minded eating can derail an otherwise well-thought-out metabolism-raising regime. Australian sports dietitian Trent Watson, Ph.D., warns, "It's well known that the major limitation to any dietary assessment is under-reporting what we eat. And snack foods seem to be the part of our diet we under-report because it's between our main meals; it's not considered a major part of what we're eating."

Extra calories can pop up anywhere—free doughnuts in the office, free samples at the grocery store, and lollipops while you're waiting in line at the bank are examples of food you probably don't factor into your daily meal plan. The health industry calls it "eating disinhibition" because you tend to let your inhibitions down in these circumstances and eat without much awareness of what you're putting in your mouth.

Subconscious Factors that Affect Overeating

Various social factors, such as palatability, the variety of food being offered, and the eating habits of the people around you all affect overeating, as confirmed by a 2005 paper published in the journal *Physiology & Behavior*. If you're presented with an array of doughnuts with different toppings and everyone else is eating them, you're more likely to take one yourself.

Studies confirm that we often eat more than we should when large quantities are available. A study published in 2006 in the *American Journal of Preventive Medicine*, showed that people take more food from larger serving bowls and when using large spoons. Another article, published in 2004 in the *Journal of Nutrition*, found that young adults who were served large portions ate significantly more than when they were served small portions of the same food.

Add Up the Damage of Unplanned Snacks

Suppose you're eating a 1,400-calorie-a-day diet, but you helped yourself to a muffin at the office (200 to 400 calories), a handful of M&Ms from your co-worker's candy dish (200 calories), a cookie and a juice box at your son's soccer game (280 calories), and a couple of mini egg rolls that were being sampled at the supermarket (60 calories). You've just added about 850 calories to your daily total, most of it through simple carbohydrates, the worst kind for burning body fat. Over the course of a month, that adds up to an extra 25,000 calories, or seven pounds (3.2 kg).

One of the simplest ways to save your metabolism from this sort of pitfall is to ensure that you're not hungry between meals. Try to make sure that each meal and snack you eat contains a complex carbohydrate, protein, and fiber. Create a plan for avoiding treats with the same diligence that you use when writing your daily meal plan. For example, knowing that you have fresh fruit or trail mix in your desk drawer can help you walk right past that box of doughnuts on the office meeting table.

The same rules apply at the grocery store, sports events, and parties. Eating at social events, particularly when there's no financial cost involved, tends to occur without much thought to whether your body actually needs nourishment. Before you pop a snack in your mouth, stop and think: Is this a food that will help me lose weight? Am I even hungry? If the answer is no, walk on by. Keeping a food journal (▶61) can help you spot those sneaky calories you barely noticed you were consuming.

83

Minimize the Metabolic Impact of Stress

Stress seems unavoidable in our fast-paced, modern society. Kenneth Pelletier, Ph.D., author of *Mind as Healer, Mind as Slayer*, estimates that between 80 and 90 percent of all illnesses are linked to stress. Additionally, he says, stress and anxiety-related problems account for 75 to 90 percent of all doctor visits. You may not be able to completely escape stressful situations in your life, but you can learn to handle them in ways that minimize their adverse effects.

Stress Raises Cortisol and Blood Sugar Levels

Brad J. King, nutritional researcher and author of *Awaken Your Metabolism: Your Ultimate Guide to Abundant Energy*, explains that the hormone cortisol, which is produced by your body when you are under stress, can compete with testosterone, which helps repair muscle tissue. Plentiful muscle tissue is a key factor in your overall metabolic rate—more muscle equates with a speedier metabolism.

To understand the impact of stress on your body, let's take a good look at cortisol. Frequently called the "stress hormone," cortisol is a corticosteroid hormone produced in the adrenal glands. It has an important survival role—it's part of the "fight or flight" reflex that gives your body a shot of energy and increased sensory awareness for a brief period of time in response to perceived danger. Following such an event, your body naturally relaxes to bring cortisol levels back to normal and restore regular body functions. In today's high-stress society, however, many people never really relax. Their cortisol levels remain elevated for extended periods of time, with negative side effects.

Long periods of increased cortisol levels raise blood sugar levels and blood pressure. A 2004 study published in the journal *Nutritional Neuroscience* suggests a link between cortisol and leptin, a protein hormone responsible for regulating appetite (▶48). High cortisol levels are associated with insulin resistance, which can lead to weight gain and other metabolism problems.

Uncontrolled Stress Can Demolish Dietary Goals

As you may already know, stress often goes hand in hand with overeating (▶73, 87). When we're stressed out, we reach for comfort foods and eat until

our blood sugar levels rise enough to make us feel better. The euphoria that accompanies a sugar rush is temporary, but the impact on your weight and metabolism lasts much longer.

A study carried out at the Yerkes National Primate Research Center at Atlanta's Emory University, published in 2008 in *ScienceDaily*, showed that chronic exposure to stress can lead to eating high-calorie foods and subsequent weight gain. As a result of this study, Mark Wilson, Ph.D., chief of the Division of Psychobiology at Yerkes, noted that stress stimulates a desire for a high-fat, calorie-rich diet, which can lead to elevated levels of cortisol and metabolism problems.

Another study conducted at the University of Alabama at Birmingham and reported in *ScienceDaily* in 2005, found that dieting combined with external stress led to binge eating. The implication is that stress plus food deprivation produces a "hedonic deprivation state," a desire for immediate gratification and rewards, which manifests as a craving for foods that are high in sugar and fat.

Metabolic wellbeing and healthy weight maintenance can require an awareness of how stress impacts your life, and establishing lifestyle patterns that enable you to avoid or at least minimize stress. Regular exercise, relaxation techniques, and various mind-body practices (see Part VII) can help you cope with stress so that it doesn't overwhelm you and undermine your metabolism goals.

84

Don't Let Sleep Deprivation Sink Your Metabolism

Do you have trouble falling asleep—or staying asleep for a full eight hours? According to a 2008 review published in the *European Journal of Endocrinology*, sleep deprivation could be a major contributing factor to today's obesity epidemic. At night, your body repairs the damage done to it during the daytime. If you don't get enough sleep—or the right kind of sleep—your metabolism is sure to take a hit.

Sleeplessness Shuts Down Your Hormones

A good night's sleep is essential to a healthy metabolism. Without adequate sleep, studies show your endocrine system quickly suffers. The resulting hormonal disruption can lead to food cravings, metabolism imbalance, and weight gain. A 2004 University of Chicago study limited male test subjects to four hours of sleep on two consecutive nights. Just those few hours of lost sleep caused a 20 percent drop in their levels of leptin, a hormone that suppresses appetite (▶48). The men also had a 30 percent rise in their levels of ghrelin, a hormone that triggers hunger. Participants' appetites for carbohydrates increased dramatically too—33 percent when it came to sugary foods and 45 percent for salty foods, such as chips and nuts.

In another University of Chicago study, participants were prevented from getting adequate deep sleep (growth hormones are released during this sleeping phase). After three nights of sleep deprivation, their sugar-metabolizing ability diminished by nearly one-quarter. A study at Stanford University in 2004 found that individuals who slept less than eight hours a night on average were heavier than those who got adequate sleep.

Too Much Stress, Too Little Sleep

Older people frequently have trouble sleeping, but sleep deprivation now appears to be increasing among young people as well. In 2002, researchers found that less than a quarter of young adults slept eight to nine hours a night. Additional stress could be the culprit—the "closet monster" that keeps us awake at night.

Stress stimulates the body's production of the hormone cortisol, giving you a quick burst of energy in response to a real or perceived danger (▶83). But if you never have a chance to really

relax, rest, and recuperate—which is true for many modern people—and remain on edge much of the time, your cortisol levels stay high. When cortisol levels stay elevated for extended periods of time, your blood sugar and blood pressure rise as well. The result can be insulin resistance, weight gain, and other metabolic problems.

How can you reduce the amount of stress in your life and get the rest you need to keep your metabolism healthy? A 2003 study published in the journal *Sleep* found that exercising in the morning enabled people to sleep better at night. Eating a light meal in the evening can also help, because digestion speeds up your metabolism and may energize you at a time when you should be relaxing (▶ **67**). Avoiding alcohol, sugar, and various food additives can also help. And many people find that meditation, yoga, tai chi, qigong (▶ **96, 10, 11, 12**), as well as biofeedback and massage (▶ **87, 92**) can reduce stress and restore mind-body harmony.

85

Guard against Hormone Disruptors

You may be doing everything right, yet your scale's still stuck and your metabolism's messed up. In order for your metabolism to function optimally, your hormone levels must be in balance. If your hormones get out of whack, your metabolism goes haywire, too. Some natural and manmade chemicals, called endocrine disruptors (EDCs), can interfere with your hormonal system. According to the U.S. Environmental Protection Agency's administrator, Lisa P. Jackson, "Endocrine disruptors can cause lifelong health problems."

How EDCs Interfere with Your Metabolism

In a nutshell, here's how EDCs wreak havoc with your metabolism. The endocrine system consists of glands and hormones that regulate a variety of life functions. The endocrine glands (pituitary, thyroid, adrenal, thymus, pancreas, ovaries, and testes) release hormones into the bloodstream, where they circulate throughout the body performing numerous functions including growth, development, maturation, and reproduction. Endocrine-disrupting chemicals interfere with the body's normal hormonal operation by either mimicking or blocking hormones.

Numerous wildlife studies show the effects of EDCs from pesticides, paint used on boats, sewage treatment runoff, and lots of other sources. These chemicals have been found to disrupt the reproduction and development of fish in the Great Lakes, seals in the Baltic region, alligators in Florida, and many other species of birds, mammals, and aquatic life. It's also suspected that EDCs can damage the hormonal systems and fertility patterns of humans. Thyroid hormones, which play an important role in metabolic function, are among the hormones that may be affected.

From 2009, the U.S. Environmental Protection Agency will require the manufacturers of sixty-seven pesticides and other chemicals to test their products to find out if they disrupt the endocrine system. The Endocrine Disruptor Screening Program (EDSP) will be expanded over time to test all pesticides.

Endocrine Disruptors are Everywhere

The problem is, EDCs are everywhere—in our food, our water, and products we use everyday. The Natural Resources Defense Council points out, "Chemicals suspected of acting as endocrine disruptors are found in insecticides, herbicides, fumigants, and fungicides that are used in agriculture as well as in the home."

According to the U.S. National Institute for Environmental Health Sciences/ National Institutes of Health, EDCs are also present in detergents, the linings of food and beverage cans, toys, furnishings and materials treated with flame-retardants, cosmetics, and dental composites. EDCs can leach out of plastics, including plastic water bottles, teething rings and nipples on baby bottles, and hospital intravenous bags. A study of seventy-seven college students, done by the Harvard School of Public Health, found that bisphenol A (BPA), a toxic chemical endocrine-disruptor known to interfere with the

sexual development of animals, leaches from plastic bottles into drinking liquid. After drinking from the polycarbonate bottles, 69 percent of the students had high urinary BPA concentrations. Plastic bottles marked with the recycling number seven on the bottom contain BPA. In 2008, Canada banned BPA in polycarbonate baby bottles. Many state governments in the United States are considering similar actions.

EDCs also accumulate in the fat cells of animals who ingest them; therefore, you're likely to be affected if you eat fatty foods or fish that come from contaminated water.

So how can you minimize your exposure to EDCs? Here are some suggestions:
• Eat organic food whenever possible
• Don't microwave food—especially fatty foods, such as meat and dairy products—in plastic containers
• Don't leave liquids in plastic bottles in a hot place, such as your car in summertime
• Don't store food in plastic containers or plastic wrap
• Don't give children plastic teethers or soft plastic toys
• Don't use pesticides in your home or on your yard
• Don't put chemicals on your pets

86

Be Alert to the Hidden Threat of Toxins

You probably don't realize how many toxic substances you use everyday that can mess up your metabolism in myriad ways. In addition to additives in your food, the chemicals sprayed on plants, industrial waste in water, and endocrine disruptors, or EDCs (▶85), your beauty and personal cleansing products may harbor hidden threats. Cosmetics and personal-care items are among the most insidious sources of toxins. The U.S. Food and Drug Administration (FDA) does not have a toxic designation for personal care products, nor does it conduct routine safety testing of them. The European Union (EU) has banned more than 1,000 ingredients used in cosmetics, but the United States has only outlawed eight so far. California is ahead of other states in passing the California Safe Cosmetics Act of 2005, requiring manufacturers that sell over $1 million a year in personal-care products to disclose any that contain a chemical that is either a carcinogen or a reproductive or developmental toxic agent, including the phthalates DBP and di(2-ethylhexyl) phthalate (DEHP).

The Not-So-Pretty Truth about Personal Care Products

A study of lipsticks by the Campaign for Safe Cosmetics (a coalition of environmental safety groups), reported in 2007, found that 61 percent of lipsticks tested contained lead. An Environmental Working Group (EWG) search of almost 900 popular lipsticks turned up only one product line—Coastal Classic Creations—that posed no hazards. (A number of small companies, however, produce organic cosmetics that are safe.)

Mercury (or the mercury preservative thimerosal), a substance that has been shown to cause brain damage, turns up in some mascaras and eye drops. Hair dyes may contain lead, the same brain-damaging stuff that's been outlawed for decades in house paint. Petroleum byproducts are found in some moisturizers. Endocrine-disrupting parabens show up in shampoos and conditioners, toothpaste, and other products.

Many sunscreens contain oxybenzone or benzophenone-3, which can cause hormone problems. Most of those with bug repellant added contain poisonous pesticides. An investigation of nearly 1,000 sunscreens by the EWG found that not only were many potentially harmful, but only 20 percent provided adequate UVA and UVB protection.

EDCs called phthalates are a common ingredient in fragrances. A 2002 study by the Campaign for Safe Cosmetics discovered that 70 percent of the substances they tested, including top-selling fragrances, contained phthalates. A repeat study by CSC, published in *USA Today* in 2008, found that a number of manufacturers are voluntarily removing some EDCs from their products. However, the FDA still considers diethyl phthalate, the phthalate used most frequently in fragrances, as safe.

The Good, the Bad, and the Really Ugly

The list of toxic and potentially harmful personal care products is long. So how can you protect yourself and your family? Start by doing the following:

- Buy and use organic products whenever possible.
- Use essential oils instead of synthetic fragrances (▶98).
- Browse the EWG's website (www.ewg.org) to see ratings for thousands of personal care products.
- Take a look at the Campaign for Safe Cosmetics' website (www.safecosmetics.org) to learn more about which products are safe and which can be harmful.
- Browse the Environmental Health Association of Nova Scotia's online guide to less toxic products (http://lesstoxicguide.ca).

PART VII

Holistic Healing: Body-Mind Techniques
to Boost Your Metabolism

81

Use Biofeedback to Control Your Metabolism

With biofeedback, which involves using your mind to control a range of bodily functions, you can augment your metabolism without dieting or working out. It's a case of "mind over matter." That might sound too good to be true, but if you think you can't influence your body's involuntary functions, think again. A 2001 paper published in the journal *Applied Psychophysiology and Biofeedback* demonstrated that training via biofeedback could, for example, allow participants to exert a degree of control over their baseline heart rate while exercising.

Control Your Resting and Exercising Metabolism

You may not be familiar with biofeedback, but researchers have been aware of its potential benefits for decades. In 1985, the American College of Sports Medicine concluded that the biofeedback "can significantly influence resting as well as exercise metabolism."

A study carried out at Vanderbilt University School of Medicine and published in 1982 in the journal *Biofeedback and Self-Regulation* found that biofeedback can aid stress-related problems with carbohydrate metabolism.

Biofeedback is regarded as a complementary healing modality, meaning it's a therapy intended to support rather than replace conventional, allopathic medicine and treatment. During biofeedback, sensors are attached to different parts of your body to measure such things as heart rate, skin temperature, blood pressure, sweat-gland activity, muscle tension, and brain activity. They might include the following, for example:

- Feedback thermometer: a thermistor or similar device attached to the finger
- Electromyograph (EMG): electrodes attached to various parts of the body; used to measure muscle tension, an indicator of headaches, migraine, and back pain
- Electroencephalograph (EEG): electrodes that measure activity in the brain
- Electrodermograph: sensors attached to the body; used to measure sweat, an indicator of anxiety
- Pneumograph: strain gauge or other device; used to measure breathing
- Photoplethysmograph (PPG): pulse oximeter; used to measure heart rate and blood flow

Having been made aware of these functions and their links to brain activity, you are then given opportunities to control them with your mind, and to monitor the results.

Use Biofeedback Techniques to Modify Physical Responses

Biofeedback enables you to understand your own metabolism so you can work to modify your physical responses and reactions. One way to do this is by controlling oxygen and carbon dioxide flow with your breathing (▶9).

Maintaining the correct balance of oxygen and carbon dioxide can elevate your metabolism, regulate blood flow, improve digestion, and enable you to exercise more effectively. According to Christopher Guerriero, founder and CEO of the National Metabolic and Longevity Research Center, "The more oxygen you take in, the harder your muscles can work, the greater their growth and development, and ultimately the faster and more efficient your metabolism becomes."

Regulating temperature and heart rate can help counteract stress and anxiety. Reduced stress levels correlate to reduced cortisol output, and excess cortisol may be linked to weight gain and slower metabolism (▶83). Biofeedback enables you to use your mind to control stress, as well as heart rate and body temperature, using mental practices such as visualization, color imagery, sound, and verbal cues, sometimes in connection with breathing techniques.

Biofeedback training is offered at clinics and hospitals around the world, with sessions being led by experienced biofeedback therapists. Once you've developed the techniques necessary to regulate your system, you can practice biofeedback anywhere and anytime, without using any special equipment. The Biofeedback Certification Institute of America (www.bcia.org) is a good place to get additional information.

88

Strengthen Your Metabolism Goals with EFT

Do you sometimes eat for reasons other than hunger? If so, you're not alone. According to a 1983 article in the *Journal of Behavioral Medicine*, studies show that "emotional eating," or eating in response to stress, anxiety, and other emotional cues, is a major cause of weight gain and can be extremely difficult to resolve. Emotional Freedom Techniques (EFT), sometimes called "emotional acupuncture," focus on resolving the emotional issues associated with food cravings and binge eating that can sabotage your metabolism-boosting goals.

EFT Aids Weight Loss

As metabolism-boosters know, sugar consumption leads to increased fat storage and has a devastating impact on your metabolism (▶75). Studies, such as one published in a 2007 article in the journal *PLoS One,* have demonstrated the addictive properties of sugar. EFT has been shown to reduce cravings for various addictive substances, as reported in an article published in 2009 in *Addiction Today* that describes the use of EFT to help inmates at Styal Prison in the United Kingdom.

The North American Association for the Study of Obesity points out that stress, level of activity, and eating habits are the three significant factors that affect your weight—stress is linked with excess cortisol and weight gain (▶83). In a study of stress, published in 2005 in the *Counseling and Clinical Psychology Journal*, EFT proved effective at alleviating stress quickly, with long-lasting results. Participants at an EFT workshop had their stress levels measured at intervals before, immediately following, and six months after the workshop. Even six months after EFT treatment, their stress levels remained reduced.

Release Emotions that Sabotage Your Fitness Program

Developed by Gary Craig in the 1990s, EFT draws on the work of Dr. Roger Callahan, who combined acupuncture with kinesiology into a practice known as Thought Field Therapy. This psychologically based modality, or therapy, addresses disruptions in the energy field that surrounds and envelops the human body. According to EFT theory, negative emotions adversely influence your body's energy field. Negative feelings and emotions are the result of negative experiences in the past. These old, deep-seated emotions can undermine your present efforts to lose weight, maintain a healthy exercise routine, and increase your metabolism. If, for example, you were a chubby child taunted by your schoolmates, you may continue in adulthood to hold negative images of yourself that derail your good intentions.

To reverse negative energy and turn it into positive energy, an EFT patient thinks of a particularly negative emotion while using her fingers to lightly and sequentially tap a series of acupuncture points on the body that lie along the energy meridians (▶89). The technique is thought to release negative emotions—and the energy associated with them—in order to rebalance the body's energy field. Often the tapping is done while you make positive statements aloud; the tapping reinforces what you say. According to some practitioners, the tapping stimulates mechanoreceptors in the skin and may increase secretion of the hormone serotonin, which is associated with positive feelings.

Note that EFT is not a proven scientific method and may not work for everyone. Although some studies, such as those cited above, have shown EFT

to have measurable physiological or psychological benefits, critics have suggested that EFT's purported benefits come from its ability to distract participants from negative thoughts. Because EFT is a low-risk, non-intrusive therapy you can do yourself, quickly and easily, it may be worth trying. As

always, talk to your physician before undertaking this therapy, especially if you're already receiving medication or other interventions. If you think you're suffering from a more chronic condition of depression or anxiety, consult with medical health professional for a complete evaluation.

Use Acupuncture to Help You Get Back in the Game

Is pain keeping you from exercising as hard or as long as you'd like to? If so, acupuncture may be able to provide relief and eliminate the need for pain-killing drugs. The World Health Organization lists more than forty conditions for which acupuncture can be used effectively, including back and shoulder pain, arthritis, tennis elbow, mobility problems, and many other health issues that could sideline you.

Acupuncture Relieves Pain and Increases Energy

According to North Carolina acupuncturists Lindsey Seigle and Brian Kramer, "Most often people express concerns about a lack of energy, increased stress, and difficulty with pain from new exercise regimens. Acupuncture is mainly helpful in weight loss because it increases patients' energy, making them want to exercise, and also alleviates the pain associated with a new exercise routine."

How can turning yourself into a human pincushion alleviate pain? Evidence from the U.S. National Institutes of Health's National Center for Complementary and Alternative Medicine shows that activating acupuncture points can speed the rate at which your system relays electromagnetic signals, increasing the flow of natural healing and pain-killing chemicals to injured parts of the body.

Acupuncture Facilitates Weight Loss

A 1993 article published in the *Journal of Traditional Chinese Medicine* discusses acupuncture's application as an aid to weight loss and weight management.

Acupuncture points associated with blood and oxygen circulation can be stimulated to improve digestion. Activating points relating to the stomach, adrenal system, and thyroid is more specifically geared toward regulating metabolism. Some doctors report that many people undergoing acupuncture, whether for metabolism issues or not, lose weight more easily while receiving acupuncture. "Acupuncture helps if there is a metabolic condition that may result in weight gain," explain Seigle and Kramer. "It increases metabolism, reduces food cravings, and improves body-wide circulation."

Providing Relief for Millions of People

Acupuncture may seem weird to some people, but more than eight million adults in the United States have used it, according to a 2002 National Health Interview Survey. Acceptance of alternative therapies is growing rapidly in the West—37 percent of hospitals in the United States now offer a variety of complementary healing modalities including acupuncture, a 2008 survey done by the American Hospital Association's Health Forum reported.

Acupuncture is a component of traditional Chinese medicine and has been in use for thousands of years. Chinese medicine proposes that health problems arise when the body's life energy (known as *chi* or *qi*) becomes imbalanced.

Acupuncture needles open blockages along energy pathways in the body, known as meridians, in order to provide pain relief and restore balance to your entire system.

Acupuncture involves the insertion of very fine needles into specific points on your body—there are more than 2,000 of them—that lie along the meridians. A treatment may require the application of dozens of needles, which are left in place for half an hour or so while the patient rests. The process isn't pain-free, but the slight pricking or pinching sensation usually lasts only a few moments.

Acupuncture generally requires more than one or two treatments to produce a continued effect. On its own it won't raise your metabolism permanently or cause you to lose weight in the absence of diet and exercise. However, it could be a valuable complementary treatment that can reduce pain and boost vitality, enabling you to improve your workouts and heighten your metabolism.

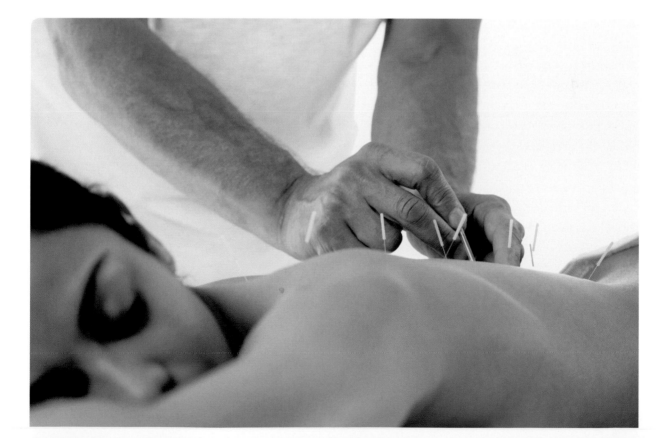

90

Employ Acupressure to Remove Energy Blockages that Slow Metabolism

If you cringe at the idea of having needles poked into your body (▶89), acupressure can provide acupuncture's benefits more gently. According to a 2005 National Institutes of Health study published in the journal *Focus on Alternative and Complementary Therapies*, acupressure is a painless, cost-free aid to weight loss. For six months, the study followed a group of patients who had lost weight. At the end of the six-month period, the subset that applied acupressure techniques kept off an average of over a kilogram (2.2 lb) more than their counterparts who didn't use acupressure.

Weight-Loss Benefits at Your Fingertips
Another study, conducted at Yuanpei University at Hsinchu in Taiwan, and published in 2007 in *Medical Acupuncture*, found that just ten minutes of acupressure once a week for eight weeks helped overweight young men lose weight. At the end of the trial, the men who used acupressure showed significantly reduced BMI, whereas the BMI for those in the control group actually increased markedly.

Like acupuncture, acupressure activates specific "control points" along the body's energy pathways, or meridians. Instead of inserting needles into these points, however, you use your fingers or thumbs to put pressure on them. Usually you apply a steady, firm hold for a minute or two, then release.

Treatment should be repeated period-ically throughout the day. Although some people experience immediate results, it may be necessary to continue to use acupressure over a span of several days, weeks, or even months to see permanent change.

The various points on the meridians cor-respond to different parts of the body. Points in the ears, for instance, may help control appetite. A number of points that influence digestion can aid weight loss, too. These include points located on the inside of the leg between the knee and the ankle. "Many overweight people retain water easily," explains Michael Reed Gach, Ph.D., author of *Acupressure Potent Points*. These points "facilitate weight loss by stimulating the metabolism to shed excess water." Acupressure points located near the elbows may help to balance digestive activities. Points between the first and

second toes may be useful for relieving stress, which can present a barrier to weight loss (▶**83, 87**).

The Tapas Acupressure Technique

A variation on traditional acupressure, known as the Tapas Acupressure Technique (TAT), combines acupressure with visualization or guided meditation. While gently pressing specific points on the face or the back of the head, you focus mentally on an issue you wish to address, such as losing weight or elevating metabolism. The practice helps remove emotional associations that may be interfering with your fitness goals. (The National Institutes of Health weight loss study referred to above used TAT.)

Eric B. Robins, M.D., co-author of *Your Hands Can Heal You*, says, "Unresolved emotional issues are stored in the body and can have a profound negative impact on healing. TAT is a powerful modality to clear negative emotions and past traumas at a body level."

In addition to being painless, one of the major advantages of acupressure is you can practice it yourself, quickly and easily. Both acupressure and acupuncture work to balance your entire system—body and mind—by removing energy blockages that can slow your metabolism. You can perform acupressure on yourself (or someone else) anytime, anywhere in order to stimulate energy, curb appetite, or relieve pain as part of your overall metabolism-boosting regimen.

91

Relieve Sports-Related and Chronic Pain with Reflexology

When back pain strikes, your exercise routine goes out the window, sometimes for months on end. At one time or another, almost everyone experiences back pain—in the United States, for example, it's the second most common neurological problem, according to the National Institute of Neurological Disorders and Stroke, costing Americans more than $50 billion annually.

A study conducted at the Hospital of Beijing College of Languages in China showed that a form of therapeutic touch known as reflexology (also called reflexotherapy) can alleviate back pain quickly, safely, and inexpensively. Researchers investigated the effect of reflexology on acute lower back pain and discovered that after only a single treatment 25 percent of patients in the study said their pain had disappeared. Another 50 percent reported being pain-free after three to four treatments, and the final 25 percent obtained relief after five to seven treatments—without using any other medications or therapies.

Foot Massage Reduces Pain

Baggage handlers in Scandinavian Airlines' cargo department (about sixty people) regularly suffered from work-related back injuries. After the airline hired a reflexologist to treat the employees, the company reported in *Reflexions*, the journal of the Association of Reflexologists, in 1993, that absentee-ism decreased substantially.

Reflexology involves massaging numerous spots on your feet and sometimes on your hands to produce results in corresponding parts of your body. The theory is, you have nerve endings in your hands and feet that are linked to every muscle and organ in your body. When sensitive points on your hands and feet are pressed, kneaded, or rubbed, signals are sent to the nervous system and brain, telling them to step up the flow of energy, oxygen, and blood to injured or ailing parts of your body. Massaging your hands and feet regularly removes blockages, promotes healing, and restores balance to your entire system.

Another study, presented at the 1993 China Reflexology Symposium in Beijing, showed that less than an hour of foot reflexology, combined with hand acupuncture (▶89), relieved sciatica pain in nearly 92 percent of patients studied. Reflexology also reduces the pain and stiffness of arthritis—that's good news for older people whose metabolisms tend to slow down with age, partly due to reduced levels of exercise. Studies published in a 1996 report by the China Preventive Medical Association and the Chinese Society of Reflexology found that reflexology reduced shoulder and/or knee arthritic pain in 97 percent of patients.

Reflexology Can Aid Obesity-Related Dysfunctions

Reflexology may also help people with weight problems deal with issues such as excessive or compulsive overeating and digestive disorders that can undermine their metabolism-boosting goals. Studies, such as one reported to the 1996 China Reflexology Symposium in Beijing, show that reflexology can aid a variety of dysfunctions associated with obesity. More than 90 percent of obese test subjects found that half an hour of foot reflexology, combined with ear acupuncture, given daily for twenty days, relieved their digestive problems, food cravings, fatigue, and numerous other conditions. A 1985 Russian study indicated that reflexology stimulated metabolism and improved the exercise capabilities of obese patients.

Pressing particular points on the big toe, for example, can help to regulate metabolic processes and hormone balance. Massaging spots near the ball of the foot has been shown to improve thyroid function and increase energy. Activating points in the arch of the foot can aid sugar balance and support digestive processes.

Although much of the scientific research done on reflexology's medical benefits has been conducted in Asia, its proponents say this five-thousand-year old form of therapy can provide a variety of other metabolism-related benefits, including increased circulation, improved respiration, and cleansing the body of toxins. Like acupressure (▶90), reflexology is an easy, safe, cost-free technique that requires little training to administer. It's also a relaxing and pleasant therapeutic practice you can perform on yourself—or someone else—anytime, anywhere.

92 Massage Your Metabolism into Shape

You may consider massage a relaxing luxury, something you treat yourself to at a spa while you're on vacation. However, research suggests that massage therapy can be a valuable aid to weight loss and weight management. A study undertaken at the University of Miami School of Medicine, published in the *International Journal of Neuroscience* in 2005, showed that massage therapy decreased levels of cortisol in test subjects by an average of 31 percent. Cortisol has been linked with increased appetite and fat deposition in the abdominal area.

Manage Your Weight by Managing Stress

Perhaps the best reason to get a massage is to reduce stress, which is a factor in virtually all illness (▶83). Cortisol, the so-called "stress hormone," is released during times of stress. High levels of cortisol have been associated with a slower metabolism, as reported in a 2002 article in the journal *Obesity Research*, and can lead to weight gain and other metabolic problems. Long periods of increased cortisol levels raise blood sugar levels and blood pressure.

A 2004 study published in the journal *Nutritional Neuroscience* suggests a connection between cortisol and leptin, a protein hormone that regulates appetite (▶48). And stress often goes hand in hand with overeating.

If you've ever had a massage, you know how good you feel afterward. There's a scientific reason for this. Not only does massage help lower cortisol levels, it also stimulates the release of endorphins. These natural "feel-good" compounds are produced during strenuous exercise—they're partly responsible for what athletes call a "runner's high." A 1985 study published in the journal *Physiology & Behavior*, showed that defects in your body's endorphin system may be related to obesity and eating disorders.

Maintain Muscle Health with Regular Massages

After a workout, a massage can aid the recovery of muscle tissue and keep your muscles limber. Massage promotes healthy muscles by increasing the rate of circulation. When skin and muscle tissues are manipulated, nutrients can flow more efficiently between blood and tissue cells. Improved circulation to muscle tissue may also reduce the amount of lactic acid that accumulates after exercise, and thus decrease soreness. Heightened endorphin activity, generated by massage, also helps ease the aches and pains that follow exercise. Endorphins produce analgesia and function as natural pain relievers.

Several types of massage can boost metabolism by increasing circulation and enhancing muscle performance. Swedish massage, the type most of us are familiar with, relaxes muscles and improves circulation through the application of deep pressure. It's one of the best forms of massage for raising metabolic rate. Sports massage targets muscles that get worked extensively during exercise, and can provide great benefits to serious athletes. Deep tissue massage uses strong, direct pressure to relieve chronic soreness and pain. Lymphatic drainage is a specialized type of massage that applies pressure near the lymph nodes to speed the elimination of toxins from the body. Neuromuscular massage therapy concentrates pressure on specific "trigger points" for up to thirty seconds at a time to increase blood and oxygen flow to the muscles and alleviate muscular tension.

Many health clubs have massage therapists on staff. Massage can be expensive, although some health insurance plans cover costs—especially in connection with rehabilitation after an injury. You may want to take a massage class with a friend or partner, so you can cut costs by giving each other massages.

93

Improve Your Ability to Exercise with Rolfing

Is limited range of motion preventing you from working out as effectively as you'd like to? A type of bodywork known as Rolfing may improve your ability to move freely, without pain. A 1977 study at the University of California's Department of Kinesiology, Los Angeles, demonstrated that after receiving Rolfing treatments patients could move more easily, with more energy and less fatigue, and exhibited better neuromuscular balance.

Rolfing Enhances Muscle Function

A 2005 paper published in the journal *Medical Hypotheses* suggests that fascial, or connective, tissue may be able to stiffen temporarily, as muscles do when they contract. This occurs in response to trauma or stress on the body, as the tissues tighten to provide additional strength and protection for the bones and organs. Over extended periods of time, however, this stiffening can become painful and impair your ability to move freely. By manipulating and stretching these tissues, Rolfing can improve both agility and muscle health.

Developed in the 1940s and 1950s by Dr. Ida Rolf, Rolfing is a type of deep-tissue massage that works with the body's connective tissue, called fascia, to relieve pain, stress, and problems resulting from injuries. This therapy is also known as Rolfing Structural Integration, because it attempts to restore balance and integration to the entire body. Therapists massage the soft tissues (fascia) between the skin and the muscles, bones, and organs to ease soreness, stiffness, and other types of pain. Rolfing enables the muscles to function more efficiently—and improved muscle function equates with improved metabolism.

Research done at the University of Maryland, and published in *Physical Therapy* in 1988, showed that Rolfing helped alleviate stress and improved body structure as well as neurological functioning. As a result, patients experienced increased range of motion, better posture, and enhanced energy.

Many professional athletes and sports teams have used Rolfing to keep them on top of their game, according to the Rolf Institute. Clients include Bret Saberhagen, Brian Oser, Craig Swan, Ivan Lendl, Edwin Moses, and Olympic gold medal cyclist Alexi Grewal.

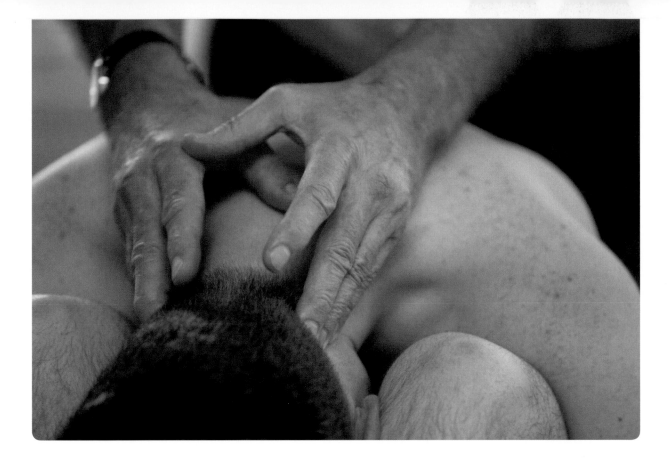

Control Chronic Pain with Rolfing

Chronic pain can bench even the most serious and stoic athletes. Like other forms of massage therapy, Rolfing has been shown to have a positive effect on many types of chronic pain. According to research published in the *Journal of Orthopedic & Sports Physical Therapy* in 1997, Rolfing offered benefits to patients suffering with chronic back pain.

Back pain often results from lack of exercise, weight gain, and weak abdominal muscles. Rolfing helps realign the back—and the entire structural system—to ease physical stress and related muscle, disc, and/or nerve pain. Once pain diminishes, patients can undertake exercise programs to boost metabolism and facilitate weight loss so their back problems don't recur. Rolfing usually involves a series of ten sessions, of about an hour each. The sessions progressively work on aligning areas of the body and connecting various layers of tissue, as well as alleviating specific discomfort or dysfunctions, so the results are cumulative and lasting.

Even though Rolfing has been practiced for more than half a century, it's just beginning to find favor in allopathic medical circles. In 2007, a conference took place at Harvard Medical School, at which doctors and medical researchers studying connective tissue interacted with practitioners who had clinical experience with Rolfing and other connective-tissue massage treatments. The conference generated huge interest, and a report of the proceedings was published in the journal *Science* in 2007. Further conferences are being planned.

94

Don't Worry, Be Happy: Laughter Boosts Your Metabolism

Did you think losing weight and raising metabolism was all about hard work, sacrifice, and pain? Think again. As it turns out, laughter may help you lose weight. A 2006 study published in the *International Journal of Obesity* showed that laughing for ten minutes a day uses about forty calories. Over a year's time, that's more than 14,000 calories. So you can actually lose four pounds (1.8 kg) a year simply by laughing! "One of the most common impediments to optimal metabolism is seriousness," says Robert K. Cooper, Ph.D., author of *Flip the Switch: Proven Strategies to Fuel Your Metabolism and Burn Fat 24 Hours a Day.*

Laugh to Lose Weight

Scientifically speaking, laughing is a physiological event that usually involves both movement and sound. When you hear or see something funny, your brain sends signals to several other parts of the body; these signals initially lead to contracting facial muscles and raising lips. The larynx closes partway, which forces an irregular (gasping) breathing pattern. If your face turns red from laughing, it's because you actually have to work to breathe while laughing due to restricted airflow.

Laughter causes the body to rev up internally. Blood pressure rises when you're laughing. Heart rate increases, and you start to breathe faster. Do these reactions sound familiar? They're the same responses your body experiences when you engage in physical exercise like running or cycling. According to the Cancer Treatment Centers of America, laughter offers a number of benefits when it comes to boosting your metabolism; it can:

- Reduce stress hormones
- Boost the circulatory system
- Enhance oxygen intake
- Trigger the release of endorphins (the body's natural painkillers)
- Improve digestion
- Balance blood pressure

Laughter Lowers Pain and Metabolism-Killing Stress

Studies, as cited in a 2003 paper in the journal *Alternative Therapies in Health and Medicine*, indicate that laughter can lower stress levels. A 1988 study at Loma Linda University School of Medicine in Southern California showed that people who watched a funny hour-long video experienced a decrease in cortisol (▶83), high levels of which have been

ankylosing spondylitis. After watching Marx Brothers' films and episodes of *Candid Camera*, he found that just ten minutes of laughter decreased his pain enough that he could sleep for at least two hours, thereby relieving the debilitating effects of sleep-deprivation-induced stress. After recovering from his "terminal" disease, Cousins helped establish a department at the University of California at Los Angeles medical school to study the relationship between illness and the mind.

Some hospitals and medical institutions now offer "laugh therapy" as a healing modality. You don't have to join a laugh-therapy group, however, to enhance your metabolism and overall health. Just spend time with people who make you laugh. Watch funny television shows or tune into radio talk shows with humorous hosts. Read comics, watch your children's antics on the playground, or enjoy stand-up comedians—do whatever makes you laugh.

linked to a slower metabolism and weight gain, as shown by a 2002 article in the journal *Obesity Research*.

Laughter is also an antidote to some types of depression. Depression, like stress, is a metabolism destroyer. But research suggests that one of the common treatments for depression—antidepressant drugs—may also be linked with metabolism problems and weight gain. A French study reported in the *Journal of Pharmacy and Pharmacology* in 2007 found that animals given antidepressants experienced a significant increase in blood glucose after six weeks of receiving the medications. Additionally, a 2005 study at the Department of Psychiatry and Psychotherapy at Philipps University in Marburg, Germany, suggested that certain psychotropic drugs may be involved in metabolic dysfunctions.

One of the best-known proponents of laughter's healing benefits is Norman Cousins, author of *The Healing Heart* (1983) and *Anatomy of an Illness* (1979). A former editor for the *Saturday Review*, Cousins decided to use laughter to relieve the crippling pain of a form of arthritis he suffered from, called

Of course, you can't realistically substitute laughter for exercise—you'd need to laugh continuously for approximately three hours to replace forty-five minutes on an elliptical trainer. But look at it this way: Each good laugh you have provides a small—but highly enjoyable—boost to your metabolism.

Overcome Metabolism-Busting Obstacles with Hypnosis

Need a little extra push to reach your fitness goals? Hypnosis offers a tool to help you overcome obstacles that may be blocking your progress. Proper weight-maintenance and exercise are the keys to raising your metabolism. Studies show that hypnosis can assist in both areas.

Using Hypnosis to Lose Weight

Weight management is one of the most popular applications for hypnosis. Multiple studies, analyzed in the *Journal of Consulting and Clinical Psychology* in 1996, compared the results of people who combined hypnosis and other weight-reduction practices with people who did not include hypnosis in their weight-loss program. Those who made use of hypnosis averaged a 97 percent increase in weight loss while undergoing treatment, and a 146 percent improvement in their weight following their treatment.

An earlier study of sixty overweight women, published in the same journal in 1986, also demonstrated the effectiveness of hypnotherapy. The group who used hypnosis lost an average of seventeen pounds (7.7 kg), whereas those who didn't only dropped an average of half a pound (0.23 kg).

Hypnosis puts into practice the idea of mind over matter. It allows you to train your mind, via the power of suggestion, to replace self-defeating behaviors with positive ones. Some researchers believe that hypnosis activates the prefrontal cortex in the brain, which is involved in attention and focus. Because you tend to be extremely suggestible during hypnosis, you can easily embed ideas and instructions into your subconscious while you're in a pleasantly relaxed, light trance state. Hypnotherapy can also help relieve stress, which can in turn increase weight loss and improve metabolic efficiency (▶83).

Modern hypnosis is a journey to a deep state of relaxation. A typical session can last anywhere from about twenty minutes to an hour. Sometimes a single session may be all that's necessary, but many people gain more benefits from a series of treatments. Although a skilled hypnotist guides the journey, you remain in control. You decide what issues you want to address—such as losing weight or breaking unwanted habits—and what goals you intend to accomplish.

Improve Your Athletic Performance with Hypnosis

Maybe the term hypnosis conjures up images of a Svengali-esque charlatan placing an unwitting subject into a zombie-like trance. But in reality, hypnosis isn't strange or scary. If you've ever daydreamed or "zoned out" while doing something, you've experienced a type of hypnotic state. Today, many athletes undertake hypnosis to improve their skills. For example, certified sports psychologist Jack Singer, Ph.D., uses hypnosis to help many professional athletes, college athletes, and Olympic competitors enhance their performance.

Hypnosis can also be an asset if you're working your way back from an injury that has restricted your ability to exercise. Numerous studies show that hypnosis speeds the rate of recovery from injuries. In 1999, for example, Carol Ginandes and Daniel Rosenthal of Harvard Medical School studied the effects of hypnosis on patients with broken ankles. The researchers found that the patients who'd been hypnotized healed in six weeks, compared with eight and a half weeks for those who had not received hypnosis.

Research demonstrates that even seemingly involuntary bodily systems, such as blood flow and heart rate, can be regulated by your mind during hypnosis. Studies done by Dr. David Spiegel, a psychiatrist at Stanford University, show hypnosis can reduce the degree of pain people experience as well. An article published in the *Journal of Pain and Symptom Management* in 1995 reported that patients experienced less pain and a lower degree of unpleasantness associated with pain as a result of the analgesic effects of hypnosis.

Hypnosis should be considered as part of a metabolism-boosting and weight-loss regime, rather than a magic cure. Although most people can be hypnotized, some exhibit a greater receptivity to trance states and consequently may achieve more satisfactory results. Make sure you use a qualified hypnotherapist.

96

Revamp Your Metabolism with Meditation

Most of us associate meditation with relaxation, not with speeding up our metabolisms. However, meditation may offer your metabolism a major boost by relieving stress, your metabolism's arch-enemy and a contributing factor to most illnesses (▶ 83). In our get-up-and-go society, few people make time in their busy schedules to meditate regularly— even though just twenty minutes a day can make a difference in your overall wellbeing. A study at Massachusetts General Hospital, reported in 2006 in *Time* magazine, showed that forty minutes of meditation per day increases your "gray matter," specifically in the brain's cerebral cortex.

Meditation Reduces Stress

A team of researchers from China and the University of Oregon conducted a study of college students to examine the effects of meditation on stress. During the five-day study, reported in a 2007 issue of *ScienceDaily,* one group received meditation training; a control group was taught other forms of relaxation. When stress was induced, the students who had meditated showed less cortisol release, as well as lower levels of anxiety, depression, anger, and fatigue.

In another study—one that Dr. Towia Libermann, director of the Beth Israel Deaconess Medical Center (BIDMC) Genomics Center in Boston, called "the first comprehensive study of how the mind can affect gene expression"— researchers at Benson-Henry Institute for Mind/Body Medicine at Massachusetts General Hospital (MGH) and BIDMC examined healthy individuals to see how meditation influenced stress response. The study, reported in *Medical News Today* in 2008, found that meditation produced "physiological changes such as in cell metabolism." These changes, explained Dr. Jeffery Dusek, co-lead author of the study, were the opposite of changes induced by post-traumatic stress disorder "and were much more pronounced in the long-term [meditation] practitioners."

Meditation and Hormonal Function

Studies, as presented in a 1986 article published in the journal *Psychosomatic Medicine,* suggest that meditation and guided imagery may improve endocrine functions and help regulate the central nervous system. This may guard against the potential effects of endocrine disruptors (▶ 85) and protect overall health.

Meditation also lowers levels of the hormone cortisol, which can compete with testosterone and interfere with the repair of metabolism-boosting muscle tissue (▶83). The adrenal glands step up production of cortisol in response to stress. Extended periods of elevated cortisol increase blood sugar levels, appetite, and insulin resistance, which can lead to weight gain and other metabolism problems. Many meditators state that regular meditation also elevates mood, alleviates chronic and acute pain, and facilitates good sleep—and lack of sleep can sink your metabolism and contribute to weight gain (▶84).

When you meditate, your brainwave frequencies drop, from 13 to 30 cycles per second (cps) to 8 to 13 cps. Heart rate and respiration slow down, too. At the same time, your brain increases production of endorphins, the proteins that enhance positive feelings. A study carried out at Emory University's Yerkes National Primate Research Center in Atlanta, Georgia, published in 2008 in *ScienceDaily*, showed that feelings of inferiority or subjugation and chronic stress led to eating high-calorie foods.

Meditation Techniques

The best-known type of meditation is Transcendental Meditation (TM); more than a million Americans practice it regularly. This technique involves repeating a "mantra" (a meaningful word, sound, or phrase) to focus your mind and silence nagging thoughts. Another technique called *progressive muscle relaxation* involves tensing, then relaxing the muscles in your body systematically, from head to toe. *Contemplation* lets you focus on a single idea, word, or question, excluding other thoughts, while allowing insights to arise in your mind—without relying on rational thinking or analysis.

Guided imagery uses a soothing image or series of images to lead your mind and body into a state of relaxation. *Creative visualization* involves creating a mental picture during meditation that realistically or symbolically represents your objective. For example, you could envision your body the way you want it to be or imagine a healing fire burning away your excess fat.

No meditative method is better than any other. A mere ten minutes of meditation a day can produce noticeable effects; however, try to work up to at least a half hour per day. If you have trouble stilling your mind—and many people do at first—try the following:

- Set aside a period of time every day to meditate—by meditating at the same time each day, you establish a routine that your mind and body will soon respond to
- Exercise lightly before meditating, to release tension that might otherwise distract you. Yoga or tai chi is preferable to kick-boxing, jogging, or sprinting (▶10, 11)
- Listen to soothing music while you meditate
- Inhale calming scents, such as lavender, while meditating, to trigger the limbic system of your brain and encourage relaxation (▶98)

225

91

Brighten Your Metabolism with Sunshine

If you feel devitalized and despondent on gray days, even when nothing obvious is wrong in your life, you're not alone. In the United States, for example, about 10 percent of the U.S. population, and more than 20 percent of those living in the northern parts of the country— perhaps as many as 35 million people— experience what's known as seasonal affective disorder, or SAD. During long winters, decreased sunlight can dampen your spirits and your metabolism. Your energy sags, hormone levels drop. Your exercise routine may take a back seat, while food cravings shift into high gear.

Sunlight Triggers Hormone Production

As noted in an article in *New Life Journal* in 2004, "Metabolism consists of the chemical processes that create energy in the body that are regulated by the endocrine system, especially the thyroid" (▶ 85). A 2001 study published in the *Journal of Clinical Endocrinology & Metabolism* examined people living in Antarctica. Researchers found that the subjects' levels of thyroid hormone dipped during the long, cold, dark winter, possibly due to lack of sunlight. Hypothyroidism (low thyroid function) is associated with weight gain, fatigue, and a sluggish metabolism (▶ 53).

A 2002 paper published in the journal *The Lancet* tied light intensity to serotonin production. Sunlight may stimulate the body to produce more serotonin, a monoamine neurotransmitter that controls mood and appetite. Interruption in serotonin function—which can result from insufficient exposure to the full spectrum of light found in sunshine—

has long been linked with depression, anxiety, low self-esteem, low vitality, eating disorders, and other emotional problems. According to Robert K. Cooper, Ph.D., author of a book on metabolism called *Flip the Switch*, "Exposure to bright light decreases melatonin and increases serotonin, shifting your body from sleep to awake mode and, in turn, revving your metabolic furnace."

The Vitamin D–Metabolism Connection

Sunlight helps the body to produce vitamin D, which it needs in order to properly absorb calcium, which helps boost metabolism (▶ 23). Prolonged sunlight deprivation decreases the body's levels of vitamin D and its ability to metabolize calcium. A study by the Naval Submarine Medical Research Lab in Groton, Connecticut, found that

seamen on patrol who didn't see the sun for sixty-nine days experienced a 42 percent drop in vitamin D.

A 2001 article published in the *Journal of the Federation of American Societies for Experimental Biology* showed that calcium also has a protective effect against the loss of lean muscle mass, a major driver of metabolism. The importance of calcium, vitamin D, and sunshine to metabolism-boosters is reiterated in a 2007 study published in the European journal *Maturitas,* which indicated that an adequate level of vitamin D may promote muscle strength.

Muscle pain may also be connected with vitamin D deficiency, according to studies published in a 2003 article in the *Mayo Clinic Proceedings*. Because vitamin D isn't present in many foods, the only ways to get it are to take supplements (or eat food supplemented with vitamin D) or spend time in the sun, without sunscreen. A 2008 article in the journal *Photochemistry and Photobiology* suggests that taking in the mid-day sun may be most beneficial. Ten to fifteen minutes is usually adequate—too much can lead to other problems. So if you're feeling devitalized or down-in-the-dumps, a brisk walk in the sun could be just the thing to shift your metabolism and your mood back into high gear.

98 To Minimize Stress, Pain, and Weight Gain, Try Aromatherapy

The scent of vanilla may remind you of baking cookies, but studies show this familiar fragrance also has powerful stress-busting properties. Stress can seriously mess up your metabolism and derail your dietary goals (▶83). In an aromatherapy study at Columbia University Medical Center in New York City, a group of test subjects sniffed vanilla while undergoing stress tests. When researchers measured the subjects' vital signs at the completion of the tests, they found the group who'd inhaled vanilla had more stable heart rates and blood pressure readings than the control group.

A Sweet Way to Soothe Stress

When you inhale a scent, it triggers your limbic system, one of the most primitive parts of the brain. The limbic system is linked with the emotions, among other things. Aromas influence brainwave function to produce both physiological and psychological responses.

"Aromatherapy is effective because it works directly on the amygdala, the brain's emotional center," explains Mehmet Oz, M.D., professor of surgery at Columbia University Medical Center. "This has important consequences because the thinking part of the brain can't inhibit the effects of the scent, meaning you feel them instantaneously."

Lavender is a scent used frequently to relieve stress and induce relaxation. A study published in a 1995 issue of *American Health* found that sniffing lavender essential oil increased alpha-wave activity in the brain. The alpha-wave frequency (about 8 to 13 cycles per second) is what your brain shifts to when you meditate, a much slower and more relaxed pattern than the ordinary, everyday beta-wave frequency. Many people find that inhaling the aroma of lavender helps them sleep better—thereby avoiding sleep deprivation, which can wreck your metabolism (▶84). A 1992 study published in the *British Journal of Medical Psychology* showed that lavender was an effective treatment for patients who experienced temporary insomnia.

Lavender helped soothe exam-time stress for student nurses at Florida Atlantic University's College of Nursing in Boca Raton, Florida. Nurses who participated in the stress study, published in *Holistic Nursing Practice* in 2009, used lavender and rosemary essential oils in sachets to relieve stress. Researchers measured the nurses' reactions and found they had lower levels of anxiety and slower pulse rates as a result of inhaling the essential oils.

Sniff Your Way to a Healthier Weight

As many weight-conscious people know, stress can lead to emotional overeating and blow your diet. Appetite and the sense of smell are closely linked. Inhaling an aroma affects the hypothalamus, an area in the brain that sends you hunger and fullness signals. Sniffing certain essential oils can cause your brain to think you're full, so you don't need to eat as much to achieve satiety.

A study carried out by the Smell & Taste Treatment and Research Foundation in Chicago, which involved more than three thousand people for a period of six months, showed that certain scents can help you lose weight. Test subjects sniffed scents including green apple, banana, and peppermint whenever they felt hungry. Without changing any other parts of their diets or exercise regimens, they lost on average five pounds (2.3 kg) per month. According to Alan Hirsch, M.D., director of the Smell & Taste Treatment and Research Foundation, "The more often people sniffed the odors, the more weight they lost."

Employing Essential Oils to Relieve Pain

Aromas can even alleviate muscle soreness after a hard workout, and ease joint pain so you can exercise more effectively. According to Dr. Hirsch, "We found that the smell of green apples reduces the severity and duration of migraine headache pain and may have a similar affect on joint pain. The scent seems to reduce muscle contractions."

A 2007 Russian study of thirty top-notch sprinters showed that aromatic essential oils had a positive effect on the athletes' motor activity. Columbia University Medical Center's Dr. Mehmet Oz found that rubbing a blend of lavender, chamomile, or eucalyptus oils diluted in a "carrier" oil, such as almond, avocado, or jojoba, expedited recovery time in his post-operative patients.

Synthetic fragrances don't produce the same effects as pure essential oils. Essential oils are extracted from plant sources, usually by distillation, and contain the medicinal properties of the plants from which they are derived. Aromatic essential oils can be dabbed on a handkerchief and inhaled throughout the day. If you prefer, add scents to bathwater, or in some cases massage essential oils directly on the skin (use caution, though, as some oils may

cause irritation). Certain oils, such as chamomile and peppermint, can be ingested. Here are some scent-sational suggestions:

- Chamomile: Reduces anxiety and aids digestion
- Jasmine: Alleviates anxiety and depression
- Lavender: Helps relieve pain, sleeplessness, and stress
- Lemon: Stimulates energy and alertness
- Orange Blossom: Relieves stress, anxiety, and insomnia
- Peppermint: Relieves pain, aids digestion, stimulates alertness
- Ylang-ylang: Alleviates anxiety, depression, and stress

229

99

Take a Hands-on Approach to Stress and Pain Relief with Reiki

If stress has thrown your metabolism off course, a soothing Japanese energy healing technique known as Reiki (pronounced *ray-kee*) may be able to bring it back into line. According to a *Natural Standard and Harvard Medical School* paper released in 2008, scientists studying Reiki found it helps relieve stress and depression, and can reduce pain as well. Researchers discovered that Reiki affects the autonomic nervous system and alters heart rate, blood pressure, and breathing activity.

Stress, Anxiety, and Depression—A SAD Combination for Your Metabolism

Anxiety, depression, and pain are often linked—and this trio can play havoc with your metabolism (▶**83**). A 2008 study at the University of Southern Maine in Kennebunk examined the

therapeutic effects of Reiki on patients over the age of sixty who suffered with pain, depression, and/or anxiety. Once a week for a period of eight weeks, Reiki masters administered half-hour treatments to the patients. Blood pressure, pulse, and pain levels were measured before and after the treatments. Although the final conclusions are still being compiled, positive patient feedback strongly indicates that Reiki significantly eased their discomfort.

Heart surgeon Mehmet Oz has worked with Reiki practitioner Julie Motz, author of *Hands of Life*, to treat patients who have received heart transplants and undergone open-heart surgery. None of the patients in the study who were treated with Reiki suffered the usual post-operative depression. Patients who

had had bypass surgery experienced significantly less post-operative pain and none of the transplant patients rejected their new organs.

Reiki has been practiced for thousands of years in Japan, China, Tibet, and other parts of Asia. The technique involves positioning the hands in special patterns to provide stress reduction, pain relief, relaxation, and other healing benefits. A Reiki practitioner may place her hands directly on your body or hold them a few inches above your body. A typical Reiki session may last from about half an hour to more than an hour, although even ten minutes can produce results.

Dr. Mikao Usui, a Buddhist monk, is credited with reviving and developing the practice in the early twentieth century. Like other Asian therapies, such as acupuncture and acupressure (▶**89, 90**), Reiki aims to balance the life force that flows through the body in order to correct disruptions that can cause illness. Philip Chan, M.D., a Reiki master in Columbus, Georgia, calls Reiki "a bioenergetic system for stress relief."

Reiki Can Relieve Pain and Get You Moving Again

It's hard to even think about exercising when you're in pain. Reiki, one of the most gentle forms of pain-relief therapy, turns out to also be one of the most effective, according to a study of 120 patients who had been in pain from a variety of causes for at least one year. The study, published in the journal *Subtle Energies and Energy Medicine* in 1998, examined Reiki as well as other forms of therapy over a period of five weeks. Reiki "proved significantly superior to other treatments on ten out of twelve variables" for pain, including the patients' total pain rating index.

In 1988, nearly 900 patients received Reiki treatments pre- and post-surgery, as reported in an article in the *Journal of Nursing Care Quality*. As a result of the Reiki sessions, patients were able to reduce the amount of pain medication they took as well as the length of time they stayed in the hospital.

According to a 2002 article in *Alternative Therapies in Health and Medicine*, more than one million adults in the United States have received Reiki for stress reduction, to decrease the need for pain medication, for sleep and appetite disorders, and to aid recovery from injuries and surgery. Reiki is now offered at more than sixty hospitals in the United States, as well as in medical clinics and hospice programs.

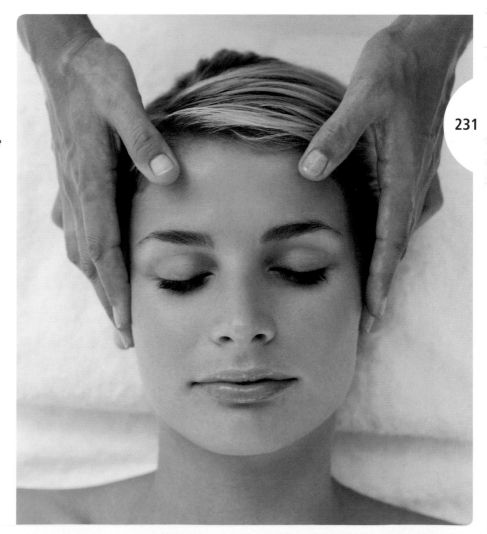

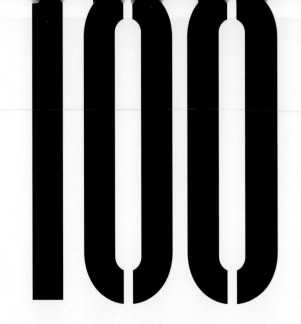

Realign Your System with Chiropractic Adjustment

When your back's out of whack, it's hard to motivate yourself to work out. Back pain (▶91) is the second most common neurological problem in the United States, according to the National Institute of Neurological Disorders and Stroke. Chiropractic adjustment is one of the fastest, most effective, and least costly treatments for back problems. Thirty out of forty-three clinical trials performed since the 1980s found that chiropractic manipulation was more effective at relieving acute and chronic back pain than other types of treatment, according to articles published in 2002 in the *Annual Review of Internal Medicine* and *Topics in Clinical Chiropractic*.

Not only back pain, but neck, hip, and joint problems can benefit from chiropractic manipulation. When your spine "locks up" or your joints aren't working properly, your muscles, tendons, and ligaments can't function optimally. As a result, you may experience decreased muscle strength and range of motion, lower endurance, and diminished performance levels. All that adds up to a slower metabolism.

Chiropractic Helps You Stay in Shape and Recover Fast

In a year-long study, published in the *British Medical Journal* in 2003, 183 patients suffering with neck pain were treated with three different types of therapy. Clinical outcome measures showed that those who received chiropractic adjustment recovered significantly faster than those who were given drugs, exercise therapy, and other forms of treatment.

Chiropractic manipulation can even alleviate the need for surgery in many cases, thereby preventing the additional trauma and downtime required for recovery. During hearings held in 1974 by a Congressional Committee regarding unnecessary surgery, the House Subcommittee on Oversight and Investigations estimated that the total number of unnecessary back surgeries performed each year in the United States could approach 44,000.

Chiropractic manipulation may also help reduce the stress that can accompany pain. Stress can wreck your metabolism (▶83) and lead to a host of other health problems. A study done at New York Chiropractic College and published in the *Journal of Manipulative and Physiological Therapeutics* in 2002, examined cortisol levels in patients who received chiropractic adjustments. Researchers found that cortisol levels decreased progressively fifteen, thirty, and sixty minutes after treatments. This is significant because the body steps up its production of cortisol in response to stress, and high cortisol levels raise the body's blood sugar levels and blood pressure, and are associated with insulin resistance, which can lead to weight gain and other metabolism problems.

Maintain Your Competitive Edge with Chiropractic Care

"Much like a high performance race car, the human body must be in proper alignment to operate at its peak performance," explain Dr. Tim McKay and Dr. Kent Jenkins, chiropractors in Calgary, Alberta. "Regular chiropractic care keeps the body functioning at optimal capacity by maintaining proper spinal alignment, which helps to eliminate the biomechanical and musculoskeletal factors that often lead to injury."

Many professional and Olympic athletes regularly receive chiropractic adjustments in order to stay on top of their game. Boxer Evander Holyfield, golfer Tiger Woods, and U.S. footballer Joe Montana are just a few noted athletes who have sought chiropractic care to improve performance and prevent injuries.

According to a 2002 article published in the *Annals of Internal Medicine*, the use of chiropractic care in the United States has increased more than threefold during the past twenty years.

Chiropractic is the best-recognized form of complementary medicine—the U.S. Congress has even included chiropractic care in its soldiers' health care coverage package and the therapy is now available at forty-two military treatment facilities nationwide.

Most people only consider going to a chiropractor when they're in pain and want immediate relief. Sometimes a single session can correct the problem and put things right again. Proponents of chiropractic care, however, recommend undergoing regular manipulations to keep your skeletal and muscular systems in good shape. Competitive athletes, in particular, and people who work out hard and want to stay in peak condition can benefit from an ongoing program of chiropractic adjustments.

Index